Digital Image Denoising in MATLAB

Digital Image Denoising in MATLAB

Chi-Wah Kok and Wing-Shan Tam
Canaan Semiconductor Pty Ltd
Adelaide, Australia

This edition first published 2024.

Registered Offices
John Wiley & Sons, Inc., 111 River Street, Hoboken, NJ 07030, USA
John Wiley & Sons Ltd, The Atrium, Southern Gate, Chichester, West Sussex, PO19 8SQ, UK

For details of our global editorial offices, customer services, and more information about Wiley products visit us at www.wiley.com.

Wiley also publishes its books in a variety of electronic formats and by print-on-demand. Some content that appears in standard print versions of this book may not be available in other formats.

Library of Congress Cataloging-in-Publication Data:

Names: Kok, Chi-Wah, author. | Tam, Wing-Shan, author.
Title: Digital image denoising in MATLAB / Chi-Wah Kok, Wing-Shan Tam.
Description: First edition. | Hoboken, NJ : Wiley, 2024. | Includes bibliographical references and index.
Identifiers: LCCN 2023055005 (print) | LCCN 2023055006 (ebook) | ISBN 9781119617693 (cloth) | ISBN 9781119617754 (adobe pdf) | ISBN 9781119617730 (epub)
Subjects: LCSH: MATLAB. | Image processing–Digital techniques.
Classification: LCC TA345.5.M42 K65 2024 (print) | LCC TA345.5.M42 (ebook) | DDC 006.6–dc23/eng/20240527
LC record available at https://lccn.loc.gov/2023055005
LC ebook record available at https://lccn.loc.gov/2023055006

Cover Design: Wiley
Cover Image: © Yuen Wai Lan

Set in 9.5/12.5pt STIXTwoText by Straive, Chennai, India
Printed and bound by CPI Group (UK) Ltd, Croydon, CR0 4YY

C9781119617693_030624

To my love, Annie, from Ted, for putting up over and over again
To my grandmother, Shui King, from Shan

Contents

About the Authors

Chi-Wah Kok was born in Hong Kong. He was granted with a Ph.D. degree from the University of Wisconsin Madison. Since 1992, he has been working with various semiconductor companies, research institutions, and universities, which include AT&T Labs Research, Holmdel, SONY U.S. Research Labs, Stanford University, Hong Kong University of Science and Technology, Hong Kong Polytechnic University, City University of Hong Kong, Lattice Semiconductor, etc. He founded Canaan Semiconductor Pty Ltd. in Adelaide, South Australia, a fabless IC company with products in mixed-signal IC, high-performance audio amplifier, high-power MOSFETs, and IGBTs. Dr. Kok embraces new technologies to meet the fast-changing market requirements. He has extensively applied signal processing techniques to improve the circuit topologies, designs, and fabrication technologies within Canaan. This includes the application of semidefinite programming to circuit design optimization, abstract algebra in switched capacitor circuit topologies, nonlinear optimization method to optimize high-voltage MOSFET layout and fabrication. He was MPEG (MPEG 4) and JPEG (JPEG 2000) standards committee member. He is an Associate Editor of Digital Signal Processing, Elsevier since 2018, and is the founding Editor-in-Chief of the journal Solid State Electronics Letters since 2017. He also is the author of four books by Prentice Hall and Wiley-IEEE, and has written numerous papers on digital signal processing, multimedia signal processing, and CMOS circuits, devices, fabrication process, and reliability.

Wing-Shan Tam was born in Hong Kong. She received her Ph.D. degree in Electronic Engineering from the City University of Hong Kong. She has been working in different telecommunication and semiconductor companies since 2004 and is currently the Engineering Manager and co-founder of Canaan Semiconductor Pty Ltd. in Adelaide, South Australia, where she works on both advance CMOS sensor design, and high-power device structure and process development. Dr. Tam has participated in professional services actively, in which she has been researcher in different universities since 2007. She has been an invited speaker for different talks and seminars in numerous international conferences and renowned universities. She has served as Guest Editor in several journals published by IEEE and Elsevier. She has been the founding editor of the journal *Solid State Electronics Letters* since 2017. She is a co-author of Wiley-IEEE textbooks, and research papers with award quality. Her research interests include image interpolation algorithm, color enhancement algorithm, mixed-signal integrated circuit design for data conversion and power management, device fabrication process, and new device structure development.

Preface

This is the second book of an ambitious project on digital image processing using MATLAB. The first book "Digital Image Interpolation" was published in February 2019. Then it is this book, "Digital Image Denoising." The last one of this project is "Hollywood Image Processing," which we are looking forward to finish writing it before the end of 2025.

The purpose of this book is to take a step further in digital image processing. Instead of considering the spatial and spectral domain to process an image, as that in the first book, this book discusses functional representation, and optimization in image processing through a practical example, "Image Denoising." Instead of suppressing undesirable signals created by processing the image artificially, the problem considered in this book is to suppress noise in image captured by modern electronic devices. The analytical tools developed in this book will be applicable to other image processing problems, as well as signal processing problems, such as pattern recognition and communication.

The book starts with discussions on noise removal through filtering, which include frequency domain-based filtering, such as mean filter and Wiener filter; and spatial domain-based rank order filtering, such as median filtering. The importance of filtering threshold selection is explained with the application of the general orthogonal transform (which include Fourier transform and Wiener filter, and hence spectral filtering as special cases). Discussions on adaptive window/ block size are presented for both spectral and spatial filtering-based denoising techniques.

The moving window filtering-based denoising techniques are extended to block transform-based denoising, and further extended to time-frequency packet-based wavelet transform in the wavelet chapter. Cycle spinning technique is applied to improve the robustness of the image denoising algorithm. Important mathematical relation between wavelet space representation and functional representation in Fourier space will be discussed.

The point-based denoising algorithm is extended to functional optimization problem in the subsequent chapters. The first set of techniques considered in this book are the low-rank matrix completion, approximation, and optimization methods. The image denoise problem is formulated as a functional optimization. However, when seeking solution to this optimization problem, it has been shown that the optimal solution is given by simple hard thresholding of the singular value. The first functional optimization-based image denoising method presented in this book is the set of variational image denoising methods. Image denoising can be achieved by variational minimization that mixes a fit to the data and the prior. In this chapter, we shall discuss the Rudin–Osher–Fatemi (ROF) total variation image denoising method, and construct the MATLAB implementation to seek the solution to this problem.

Intuitively, the image processing problem can be formulated as "deterministic" or "probabilistic" problems. Techniques presented in initial chapter have treated the image denoising problem using deterministic signal model. The last two chapters of the book will discuss techniques that make use of the probabilistic signal model. We start with the patch-based image denoising, the NonLocal mean image denoising method. With limited space and the objective of presenting algorithms that are generally useful across various signal processing field, we have made our choice to discuss patch-based image denoising with self-similarity that does not need an added dictionary, nor a prior training.

The last chapter will discuss the application of random sampling to mix and match various denoising algorithms together to achieve better result. It also demonstrates how to mix and match signal processing techniques developed with different models and in different signal spaces to achieve a better denoising result.

All discussions are accompanied with a thorough discussion on MATLAB implementation, where source codes are provided and embedded into the text as part of the discussions, and explanation on the difficult mathematics. A unified set of test images is applied throughout the whole book to allow reader to easily appreciate, compare, and observe the pros and cons of various discussed algorithms. The noise model being considered are additive white Gaussian, and Salt and Pepper (Poisson) noise, which are noises commonly found in modern digital camera photos. Their characteristics, generation, and visual appearance will be presented in the first chapter, where the first chapter will also help to warm up the reader by introducing notations being applied in the book, together with some basic MATLAB programming techniques for image processing. It will also present image quality metric and their development in MATLAB, and other mathematical and MATLAB tools that are required in the latter chapters. Only one image will be used throughout the book to provide consistency and ease of comparison. The chosen image is the "*Sculpture*" image on the front cover of the book. This is a very well-crafted sculpture located at the Hong Kong Museum of

Art and was captured by Miss. W. L. Yuen who provided us permission to use it in this book. Besides being an excellent photo, this image contains important features that have made explaining image denoising in this book easier. At the same time, we encourage the readers to experience the performance of algorithm presented in this book and also algorithm developed by themselves after reading each chapter with other images of their choice.

January 2024
Adelaide, Australia

Chi-Wah Kok, Wing-Shan Tam

Acknowledgments

I would like to express my profound gratitude to my wife, Annie. After a long day of work, I find my reward is to have your head lying on my shoulder with the satisfaction of feeling comfortable and safe. I thank my beautiful and intelligent wife, with whom I can share this and so many other things, whose love and support through the years have had an immeasurable impact on my life. I sincerely believe that she deserves much more than what I can express with my words. She is always the beautiful music in my heart.

I am fortunate to see the transformation of Dr. Tam after a very long apprenticeship instead of a clone of myself. Dr. Tam has a transformed mind, a transformed way of thinking, talking, performing, arguing, writing, and even a transformed way of walking. The idea of apprenticeship is to share work, discover new theories together, but with the immense benefit of the master's experience being challenged by the apprentice's fearless questioning. Dr. Tam's sharp and judicious remarks greatly helped me to better describe many of the ideas found in this book. Dr. Tam is always the voice of challenge in my curiosity cabinet of creativity.

> Take away from Me the noise of your songs; I will not even listen to the sound of your harps.
>
> – Amos 5:23 (NIV)

Chi-Wah Kok

The "transformation" is a process to reshape an object or a collection of objects in between different planes or coordinate systems. There are different types of "transformation," e.g. reflection, rotation, shearing, etc., which are common operations in image processing. Different types of transformation can be applied independently or multiple of them can be invoked simultaneously to achieve the desired effect. No matter what type of transformation that we are talking about, we can formulate a well-defined mathematical function to describe and to direct such transformation. There would be no ambiguous and unexpected results in the course of the transformation in the mathematical world. However, this might not be the case in the human world.

In human perspective, a transformation is sometimes unexpected, and the outcome is almost not predictable. The transformation requires courage brought by encouragement, persistence fostered by immaculate caring and understanding, and insight ignited by unreserved guidance. I am fortunate enough to experience an amazing transformation and gaining much power and strength from it. I would like to express my gratitude to everyone who nurtured my transformation. I would especially like to thank my parents for raising me up with full of encouragement, caring, and love. I would also like to thank my master, the co-author of this book, Dr. Kok, for his inspiring and endless guidance. All the tangible and intangible support from the people surrounding me have paved the way for my transformation and it will continue.

I do not take the opportunity for writing my third book for granted. I cherish the opportunity and trust this is part of the plan from God, reminding me to be humble to learn, to be rigorous to write, to be grateful for all opinions, and to be joyful to share the Good News.

> Praise the LORD, for the LORD is good; sing praise to his name, for that is pleasant.
>
> – Psalm 135:3 (NIV)

Wing-Shan Tam

Nomenclature

1D	one-dimensional
2D	two-dimensional
ADC	analog-to-digital converter (A/D)
AWGN	additive white Gaussian noise
CCD	charge-coupled device
CMOS	complementary metal-oxide semiconductor
CFA	color filter array
dB	decibel
DCT	discrete cosine transform
DFT	discrete Fourier transform
DSLR	digital single lens reflex
DTFT	discrete time Fourier transform
DWT	discrete wavelet transform
FFT	fast Fourier transform
FIR	finite impulse response
FRIQ	full reference image quality index
HR	high-resolution
HVS	human visual system, describing how humans perceive and interpret visual images
IID	independent and identically distributed
IDCT	inverse discrete cosine transform
IDFT	inverse discrete Fourier transform
IDWT	inverse discrete wavelet transform
IIR	infinite impulse response
JPEG	joint photographic experts group
LPF	lowpass filter
LR	low-resolution
MATLAB	high-level technical computing language by MathWorks Inc.
MAD	median absolute deviation

MED	median
MOS	mean opinion score
MSE	mean squares error
MSSIM	mean structural similarity
NRIQ	no reference image quality index
PDF	probability density function
PSNR	peak signal-to-noise ratio
RGB	red, green, and blue color space
RMSE	root mean squares error
SAP	salt and pepper noise
SNR	signal-to-noise ratio
SSIM	structural similarity
SVD	singular value decomposition
WSS	wide-sense stationary
WT	wavelet transform
YCbCr	luminance, blue chrominance, red chrominance color space

$\lceil x \rceil$	Ceiling operator that returns the largest integer lesser than or equal to x
$\mathbb{Z}$	the set of integer
$\mathbb{Z}^+$	the set of positive integer (greater than 0)
$\mathbb{R}$	the set of real number
$\mathbb{C}$	the set of complex number
$(\mathbf{x})$	vector x defined in continuous domain
$[\mathbf{x}]$	vector x defined in discrete domain
$[a(m,n)]$	a 2D function defined on continuous (m,n) Cartesian domain
$[a[m,n]]$	a 2D function defined on discrete $[m,n]$ Cartesian domain
$\mathbf{A}_{M,N}$	arbitrary matrix of size $M \times N$ constructed by matrix entrance $a[m,n]$ with $\mathbf{A}_{M,N} = [a[m,n]]_{m,n}$
$\mathbf{I}_N$	identity matrix of size $N \times N$
ℓ_2	the space of all squares summable discrete functions/sequences
$\mathcal{L}^2$	the space of all Lesbeque squares integrable functions
$\mathcal{R}$	teal part of a number, matrix, or function.
$\mathcal{I}$	imaginary part of a number, matrix, or function.
$\mathbf{sinc}(x)$	Sinc function $\left(\frac{\sin(x)}{x}\right)$
δ	Kronecker delta, or Dirac-Delta function, or unit impulse with infinite size
W_N	N-th root of unity and equals to $e^{\frac{-j2\pi}{N}}$
F	discrete Fourier transform operator
F^{-1}	inverse discrete Fourier transform operator

$\mathbf{W}_N$	discrete Fourier transform matrix of size $N \times N$; $\mathbf{W}_N = [W_N^{k,\ell}]_{k,\ell}$. The Fourier matrix is of arbitrary size when N is missing
$\mathbb{W}_{(m,n)}$	a window that specifies a collection of pixel locations around the $[m, n]$.
$\otimes$	convolution operator
Δ_x	interval in domain x; the interval domain is arbitrary when x is missing

A word on notations

1. (*Indices*): We denote continuous variable (m) and discrete variable $[n]$ indexed signals as $x(m)$ and $x[n]$, respectively.
2. (*Vector-matrix*): The blackboard bold $(\mathbf{A})$ is used to represent matrix-valued signal and function, and $(\mathbf{x})$ is used to represent the vector-valued signal and function. The normal characters (x) are used to represent signal in scalar form.
3. (*Rows versus columns*): For vector-matrix multiplication written as $\mathbf{xA}$, we may take vector $\mathbf{x}$ as a row vector.

About the Companion Website

This book is accompanied by a companion website:

www.wiley.com/go/kokDeNoise

This website includes:

- MATLAB codes
- PowerPoint files[1]
- Solutions Manuals[1]

1 PowerPoint files and Solution manuals are available upon registration for Professors/lecturers who intend to use this book in their courses.

1

Digital Image

An image is a two-dimensional (2D) light intensity function $f(x, y)$, where (x, y) is a coordinate system of interest. Without loss of generality, and to simplify our discussions, the rest of the book will concentrate on the case of 2D Cartesian coordinate system. The value of f at the coordinates (x, y) is proportional to the brightness of the image at that point. While digital images can be generated/acquired by a number of methods, primarily, the image f is converted to a digital image through cameras using a 2D image sensor array. These sensors are typically constructed with *charge-coupled devices* (CCD) and *complementary metal oxide semiconductor* (CMOS) technologies. Camera constructed with CCD or CMOS works in a similar fashion, where the light reflected from an object will impinge onto the face of the sensor array, such that each sensor element in the array will generate an electrical signal $f[m, n]$ (for which the coordinate $[m, n]$ can be considered to be the digitized coordinate of (x, y)). Figure 1.1 illustrates the construction of a color digital camera which is used to capture the *Sculpture* image. The light bounced off the *Sculpture* will be focused onto the sensor array through the lens. Consider a sensor array with M-rows and N-columns, the output of the sensor array will be a $M \times N$ matrix $f[m, n]$ with $0 \leq m \leq M - 1$, and $0 \leq n \leq N - 1$. As a result, the arrangement of the image sensor array is also known as the *sampling grid*, where the intersection of a row and a column will be assigned with an integer coordinate $[m, n]$ in the discrete Cartesian coordinate system. The output $f[m, n]$ of each sensor element represents the number of photons that react with the sensor at location $[m, n]$. The output of the sensor array is not a digital image yet. The subsequent *analog-to-digital converter* (A/D converter) accomplishes the quantization processes of the light intensity at all $M \times N$ locations to generate the digital image. The sampled image obtained from the sampling and quantization process, as shown in Figure 1.2(a), is the discrete image which forms a matrix $[f[m, n]]$: $0 \leq m \leq M - 1, 0 \leq n \leq N - 1 \in \mathbb{Z}^+$. Each entry in this array, $f[m, n]$, records the

Digital Image Denoising in MATLAB, First Edition. Chi-Wah Kok and Wing-Shan Tam.

Companion website: www.wiley.com/go/kokDeNoise

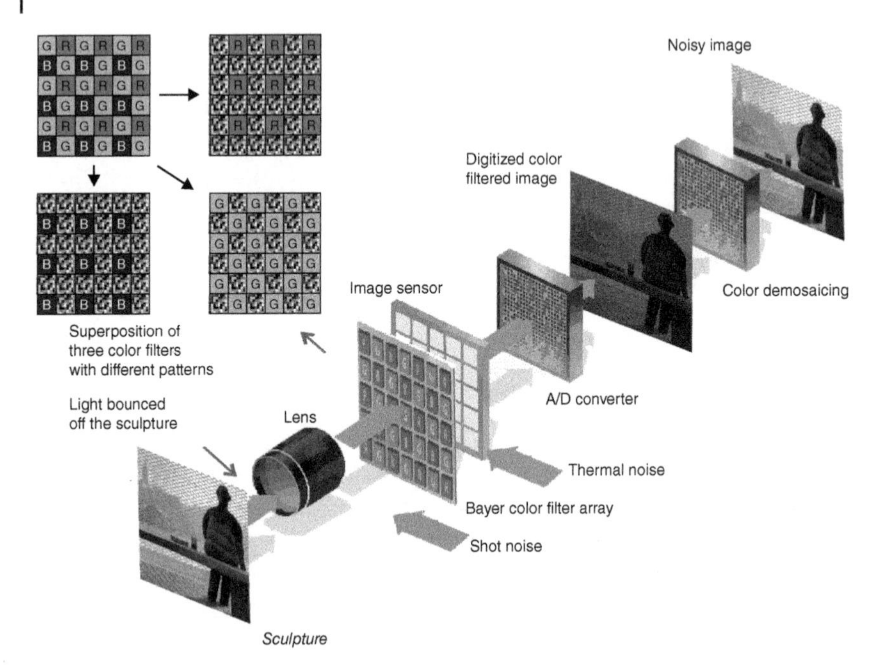

Figure 1.1 Illustration of capturing an image by digital camera.

number of photons sensed by the corresponding sensor in the arrays and is termed a *pixel*. Thus, a digital image obtained by a digital camera will look like

$$[f[m,n]] = \begin{bmatrix} f[0,0] & f[0,1] & \cdots & f[0,N-1] \\ f[1,0] & f[1,1] & \cdots & f[1,N-1] \\ \vdots & \vdots & \ddots & \vdots \\ f[M-1,0] & f[M-1,1] & \cdots & f[M-1,N-1] \end{bmatrix}. \tag{1.1}$$

The values assigned to every pixel are the brightness recorded by the image sensor, which is also interpreted as the pixel *intensity* (also known as the *gray-level* or *grayscale*).[1] To store, transmit, and visualize the discrete image, the pixel intensity of the discrete image will be rounded to the nearest integer value within L different gray levels through the quantization process performed within the A/D converter. This process will produce the digital image, which can be visualized as a shade of gray denoted as the *grayscale* or *-level* value ranging from black (0) to white ($L-1$), such that the higher the intensity value, the brighter

1 The authors do not want to join the fight between "greyscale" and "grayscale." It is however, MATLAB adopted "grayscale," and we adopted MATLAB, and thus this book will use "grayscale."

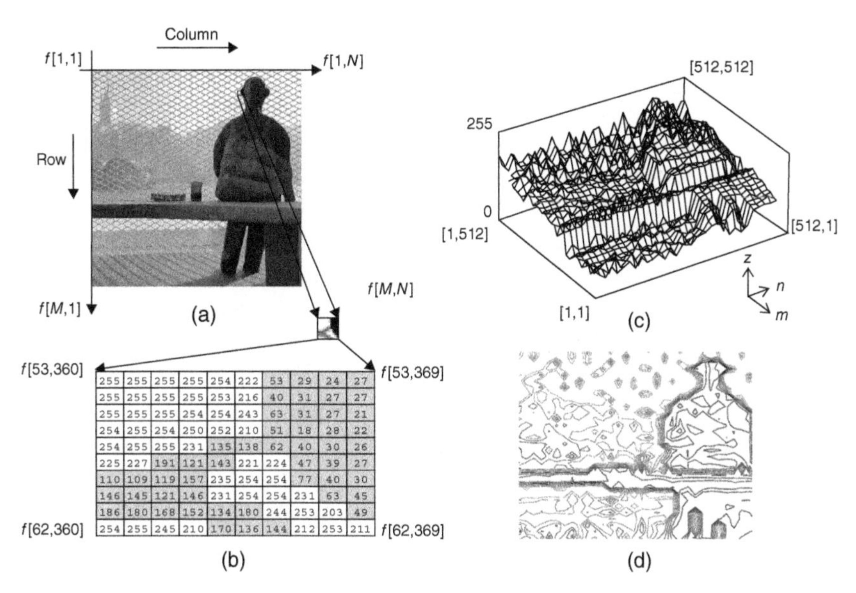

Figure 1.2 Representation of the digital image *Sculpture* : (a) a grayscale printout of *Sculpture*, which is described by an $M \times N$ 2D array within the computation system with each matrix element representing the intensity of a pixel taking a value in the quantizer (in this case, it is [0,255] as *Sculpture* is an 8-bit quantized image); (b) a pixel intensity map of the selected region in the image, where the pixel intensity at [62,369] is 211; and the intensity variation across the complete image by viewing (c) the 2D vector mesh of the image on a plane with the height (z-axis) being the pixel intensity or through (d) the contour map, where the pixel with the same intensities are located to the same contour lines.

the image pixel. Figure 1.2(b) shows the pixel values of an extract from the image $f[m,n]$.

The discrete image is arranged with each pixel $f[m,n]$ being located at the mth row and n^{th} column starting from the top-left image origin (as shown in Figure 1.2(a)) with respect to the MATLAB convention. For simplification, we shall also use the vector $p=[m,n]$ to represent the pixel location, such that $f[p]=f[m,n]$. Now, the readers may have already noticed from Figure 1.2(a) that the matrix indices in the figure are different from those in Equation 1.1. This is one of the irritating features of MATLAB. Notwithstanding the similarity between the arithmetic and the language of MATLAB, all matrices within MATLAB are indexed with the top left-hand entry as $[1,1]$ instead of $[0,0]$, and hence the discrepancy between Figure 1.2(a) and Equation 1.1. The rest of the book will assume this difference to be natural and will no longer discuss the difference between the MATLAB implementation and the analytical analysis with respect to the indexing problem.

1.1 Color Image

As pointed out by Sir Isaac Newton, color is perceived by the mind to resolve the interaction of light sources, objects, and the visual system, which adds a subjective layer on top of the underlying objective physical properties – the wavelength of the electromagnetic radiation carried by color signal. The color signal is received by light-sensitive cells in human eye. Hering's experimental results and the discovery of three different types of photosensitive molecules in human eyes [52] led us to the modern color perception theory, where color is perceived through a *luminance* (grayscale) and two *chrominance* (color) components. This is the basis of trichromacy, the ability to match any color with a mixture of three suitably chosen primaries. The basic principle of color additivity has led to a number of useful trichromatic descriptions of color, which is also known as the *color space.*

Among various color spaces, the RGB, and the YCrCb are the most popular. In particular, the RGB color space has been widely employed in digital cameras and monitors to capture and display digital color images. This is because the RGB space conveniently corresponds to the three primary colors which are mixed for display on a monitor or similar devices. A digital color image in the RGB space is similar to a digital monochrome (grayscale) image, except that it requires a three-dimensional vector to represent each pixel, and thus three $M \times N$ arrays are required to represent the whole image. Each of these $M \times N$ array represents one of the RED, GREEN, and BLUE primitive color components. The RED, GREEN, and BLUE components of an RGB image can be viewed separately as a monochrome image by considering the corresponding $M \times N$ array alone, as shown in Figure 1.3. When the three color components are superposed, it produces the rightmost color image in Figure 1.3. As a result, if each component image is encoded with the data type `uint8` in MATLAB, the total number of bits

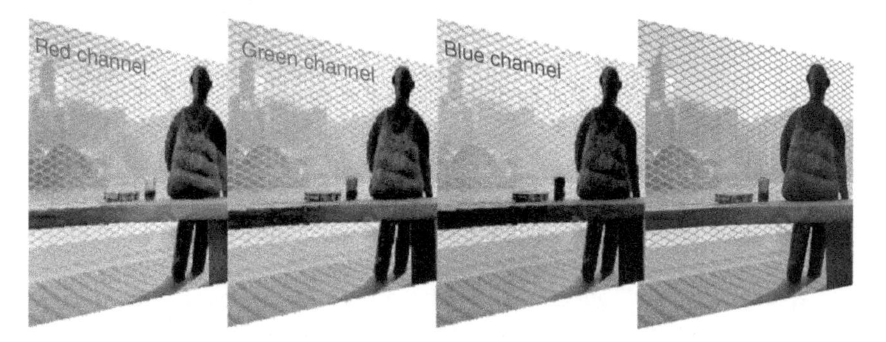

Figure 1.3 Three separate RED, GREEN, and BLUE channels are combined to create a final, full-color image.

required to represent each pixel will be 8 bits × 3 = 24 bits. This is also the default representation adopted by MATLAB for the three color triplets, and such type of image is known as the *True Color* image. Disregarding the digital color image format, the MATLAB function `imread` can be used to import the image directly from the image file stored in the hard disk, as shown in Listing 1.1.1.

Listing 1.1.1: Digital image details.

```
>> f = imread('sculpture_color.tif','tiff');
>> whos f
   Name        Size                Bytes     Class
   f           512x512x3            786432  uint8
```

Although only grayscale image denoising algorithms are discussed in this book, the algorithms can be easily extended to color images by treating the spectral components of the color images as independent grayscale images. Actually the grayscale image contains a lot of information, and this is the reason why black-and-white television receivers have been perfectly acceptable to the public for many years, and black-and-white photographs are still popular. Nevertheless, color is an important property, and so we shall examine its role in this section.

1.1.1 Color Filter Array and Demosaicing

To capture a digital image in color, three sensors with each sensor measuring one of the three colors, respectively, are required to capture the RED, GREEN, and BLUE component images. A cheaper alternative to the three-sensors camera system is to have one sensor only. In this case, each photo sensor in the sensor array is made to be sensitive to one of the three colors (ranges of wavelengths). This can be done in a number of different ways. A popular method in modern camera is to cover the photo sensor array with a *Bayer pattern* color filter array (CFA) [3], as shown in Figure 1.1. Besides the Bayer pattern CFA, the readers may have also noticed that there is a color demosaicing block by the end of the camera in Figure 1.1. These two modules are essential in capturing images with color. To be more precise, Bayer pattern CFA is a typical construction of CFA, which is commonly applied to the photo sensor arrays in modern cameras. The filter is arranged in Bayer pattern which is a combination of RED, GREEN, and BLUE filters in checkerboard format. The size of the CFA is identical to that of the sensor array and each color filter has a narrow passband, and will only allow the light component with the same color tone as that of the filter to pass through. Therefore, each pixel in the "digitized color filtered image" is the intensity of one of the three color tones of a color pixel. An example of a 6 × 6 *Bayer pattern* color filter

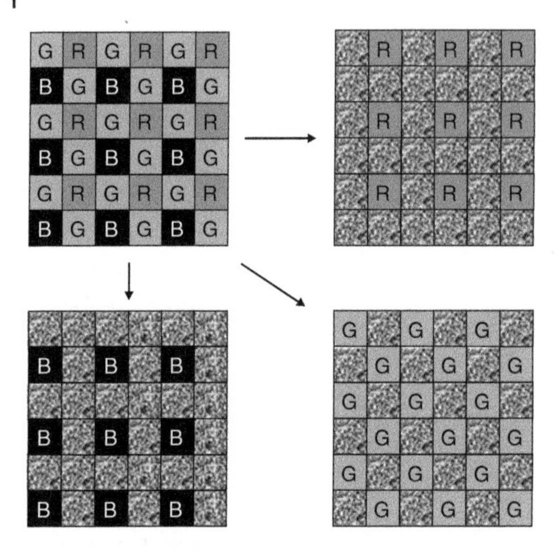

Figure 1.4 RED, GREEN, and BLUE samples obtained from *Bayer pattern* color filter.

arrangement is shown in Figure 1.1 with details in Figure 1.4, which can be separated into three images containing the RED, GREEN, and BLUE pixels separately with some undefined pixel values (as shown by the shaded box in the figure). A full-color image needs the information of all three color components in each pixel location. The color demosaicing block in Figure 1.1 takes the role to estimate a color image that faithfully resembles the real image captured by the camera by interpolating the two missing colors in each pixel location using the information of the neighboring pixels. Although complicated algorithms do exist, most color demosaicing algorithms consider each primitive color image separately, and the missing pixels in each color component are interpolated independently. The color image is obtained by superimposing the three mono-color images that contain the RED, GREEN, and BLUE pixel intensities together.

1.1.2 Perceptual Color Space

Most digital image processing algorithms make use of a simplified RGB color model (based on the CIE color standard of 1931) that is optimized and standardized toward graphical displays. However, there is a primary problem with the RGB color model, which is the RGB model is perceptually nonlinear. By this, we mean that moving in a given direction in the RGB color cube does not necessarily produce a color that is perceptually consistent with the change in each of the channels. For example, starting at white and subtracting the blue component produces yellow; similarly, starting at red and adding the blue component produces magenta. For this reason, RGB color space is inherently difficult for humans to work with, because it is not related to our perception on natural colors. As an alternative, we may use perceptual color representations, such as YCbCr color space.

Although color is typically thought of as being 3D due to the trichromatic nature of color matching, five perceptual attributes are needed for a complete specification of color appearance. These are *brightness*: the attribute according to which an area appears to be more or less intense; *lightness*: the brightness of an area relative to a similarly illuminated area that appears to be white; *colorfulness*: also known as *chromaticness*, which describes the attribute according to which an area appears to be more or less chromatic; *chroma*: the colorfulness of an area relative to a similarly illuminated area that appears to be white; *hue*: the attribute of a color denoted by its name such as blue, green, yellow, and orange. In some cases, the attribute *saturation* that describes the colorfulness of an area relative to its brightness may be provided, which can be derived from the previous five attributes, and is therefore considered redundant.

The *human visual system* (HVS) is less sensitive to chromatic than to luminance. In the RGB color space, the three primitive colors are equally important, and so they are usually all stored at the same resolution. It is possible to represent a color image more efficiently by separating the chromatic information and representing luma with a higher resolution than that of the chromas. The YCbCr color space and its variations are the more popular ways to efficiently representing color images. Y is the luminance (luma) component and can be calculated as a weighted average of R, G, and B as

$$Y = k_a[1] + k_r[1]R + k_g[1]G + k_b[1]B, \tag{1.2}$$

where the vector component k's are the weighting factors. The color information can be represented as *color difference* (chrominance or chroma) components, where each chrominance component can be calculated as a weighted average (difference) of R, G, and B with the rest of the vector components of k as

$$C_b = k_a[2] + k_r[2]R + k_g[2]G + k_b[2]B, \tag{1.3}$$

$$C_r = k_a[3] + k_r[3]R + k_g[3]G + k_b[3]B, \tag{1.4}$$

with the constant vector k_a being added to shift the range of the chrominance components to the range of $[0, 255]$ with the input RGB signal range being $[0, 255]$. In other words, each color component in a color pixel would require 8 bits, and a total of 8 bits $\times$ 3 = 24 bits are used to represent the color for each pixel in the image. The *color depth* or *color resolution* of such an image is known as 24-bit, which is the number of bits used to store the color information for each pixel. A MATLAB implementation that converts RGB to YCbCr and back is given by `rgb2ycbcr` and `ycbcr2rgb`, respectively.

Listing 1.1.2: RGB to YCbCr conversion.

```
function fg = rgb2ycbcr(f)
  ka = [16,128,128];
  kr = [65.481,-37.797,112];
  kg = [128.553,-74.203,-93.786];
  kb = [24.966,112,-18.214];

  r = im2double(f(:,:,1));
  g = im2double(f(:,:,2));
  b = im2double(f(:,:,3));

  Y  = ka(1) + kr(1).* r + kg(1).*g + kb(1).*b;
  Cb = ka(2) + kr(2).* r + kg(2).*g + kb(2).*b;
  Cr = ka(3) + kr(3).* r + kg(3).*g + kb(3).*b;
  fg(:,:,1) = Y;
  fg(:,:,2) = Cb;
  fg(:,:,3) = Cr;
  fg = uint8(fg);
end
```

The conversion can also be performed efficiently using vector-matrix multiplication, as shown in `ycbcr2rgb`.

Listing 1.1.3: YCbCr to RGB conversion.

```
function fg = ycbcr2rgb(f)
  ka = [16;128;128];
  k = [65.481,128.553,24.966;
      -37.797,-74.203,112.0;
      112.0,-93.786,-18.214];
  kt = inv(k);
  tb = kt*ka;
  Y = im2double(f(:,:,1));
  Cb = im2double(f(:,:,2));
  Cr = im2double(f(:,:,3));
  r = mat2gray(-tb(1,1) + kt(1,1).* Y + kt(1,2).*Cb + kt(1,3).*Cr);
  g = mat2gray(-tb(2,1) + kt(2,1).* Y + kt(2,2).*Cb + kt(2,3).*Cr);
  b = mat2gray(-tb(3,1) + kt(3,1).* Y + kt(3,2).*Cb + kt(3,3).*Cr);
  fg(:,:,1) = r;
  fg(:,:,2) = g;
  fg(:,:,3) = b;
  fg = im2uint8(fg);
end
```

In the case of YCbCr color space, the color image is completely described by Y (the luminance component) and C_b, C_r, and C_g (the three chroma components), where the subscripts "*b*," "*r*," and "g" denote the blue, red, and green channels, respectively. Since $C_b + C_r + C_g$ is a constant and so only two of the three chroma components are needed to be stored or transmitted, as the third component can always be calculated from the other two. As a custom, the blue and red chroma (C_b, C_r) are selected in the YCbCr color space, and C_g will be computed from these two color chroma. YCbCr color space has an important advantage over the RGB color space, where the C_r and C_b components may be represented with a *lower resolution* than that of Y because the HVS is less sensitive to chroma than luminance. This reduces the amount of data required to represent the chrominance components without introducing obvious artifacts to the image. As a result, the YCrCb color space has been adopted in a number of international image storage standard, such as JPEG. It should also be noted that each entry in the YCbCr color space is required to be an unsigned 8-bit integer to achieve the same resolution as that of an RGB color space with 8-bit per color component through `rgb2ycbcr`. In other words, the two color systems have consistent color depth to describe the color image. However, the conversion between color spaces is a non-invertible transform, such that the true color information (RGB) will be lost in the conversion and cannot be truly recovered.

1.1.3 Grayscale Image

The MATLAB function `rgb2ycbcr` has shown a way to convert a color image to a grayscale image, which is the `y` component in the `rgb2ycbcr` conversion. It is acceptable to use the Y-component alone as the grayscale image. However, it is more commonly accepted to use the mean value of the RGB components as the grayscale image. The following MATLAB function `rgb2gray` implements this grayscale image generation.

Listing 1.1.4: RGB to grayscale image conversion.

```
function g = rgb2gray(f)
  r = im2double(f(:,:,1));
  g = im2double(f(:,:,2));
  b = im2double(f(:,:,3));
  temp = (r+g+b)/3;
  fg = im2uint8(temp);
end
```

The conversion will generate an $M \times N$ array with integer values in the range of $[0, 255]$. This function will be used to generate the grayscale image from the full-color *Sculpture* image, where the grayscale *Sculpture* image will be applied throughout all denoising algorithms in this book.

1.2 Alternate Domain Image Representation

Some might notice that the digital image can be represented by other forms that are not pixel oriented. For example, the JPEG image [40] is not stored in pixel array form to achieve a compact storage size. Nonetheless, pixels are still the central concept in digital imaging. The quality of the digital image grows with the spatial, spectral, radiometric, and time resolutions of the digitization process. This image model allows us to identify the set of digital images with complex vector space $\mathcal{L}^2(\mathbb{Z}_M \times \mathbb{Z}_N)$, where complex space is used because we are preparing for the transform analysis to be used by the rest of the book:

$$\mathcal{L}^2(\mathbb{Z}_M \times \mathbb{Z}_N) = \{f : \{0, \ldots, M-1\} \times \{0, \ldots, N-1\} \rightarrow \mathbb{C}\}, \tag{1.5}$$

which f is a map. The following vector product is also defined in the complex space:

$$\langle f, g\rangle = \sum_{m=0}^{M-1} \sum_{n=0}^{N-1} f[m,n]\overline{g[m,n]}, \tag{1.6}$$

where the overline denotes complex conjugation. Note that f and g in Equation 1.6 are two matrices of the same size. Since an image is presented in this vector space through spatial coordinates $[m, n]$, the domain of this vector space is also termed as the *spatial* domain.

The row-by-column representation of the digital image defines the number of pixels used to represent the analog image, which is known as the sampling rate, and also referred to as the pixel or *spatial resolution* (also known as *geometric resolution*) of the digital image. The notation $M \times N$ is commonly used to denote the spatial resolution. The variation in intensity across the image pixels can form a vector-valued mesh on a plane, where the mesh location is defined by the pixel location. An example of the functional plot of the mesh defined by the image $f[m, n]$ in Figure 1.2(a) is shown in Figure 1.2(c). However, the spatial resolution does not tell us much about the actual appearance of the image as realized on a physical device. The *resolution density*, which gives the number of pixels per unit length, such as the *pixels per inch* (ppi) or the *dots per inch* (dpi), is the common unit that enables us to obtain the dimensions of the image. Such that when specified together with the spatial resolution, the actual image size will be determined.

It should be noted that $f[m, n]$ is a 3D discrete function. Furthermore, it is constrained to be a non-negative function. Due to the digitization process, this function will only take on a finite number of values (in our case, L different values). The number of quantization levels will affect the *radiometric resolution* of the digitized image which measures the number of distinguishable gray levels in the digital image. An effective method to display the quantization effect of the image is by means of image isophotes, which are curves of constant gray value, such that $f[m, n] = c$ for a given c, with analogy to iso-height lines on a geometric map. An image isophote plot of the *Sculpture* image in Figure 1.2(a) is shown in Figure 1.2(d).

The spatial resolution and the radiometric resolution are the two factors affecting the quality of the digital image. These two factors can be altered by users during the image capturing processes. Besides, the *spectral resolution* that specifies the bandwidth of the light frequency that can be captured by the sensor, and the *time resolution* that measures the interval between time samples at which images are captured are technology-dependent and are therefore seldom discussed in digital image processing, but rather in the semiconductor device and circuit of the image capturing devices.

1.3 Digital Imaging in MATLAB

Digital images can be stored on a computer in a number of formats, such as "`bmp`" (bit map), "`tiff`" (tagged image file format), "`jpg`" (joint photographic experts group), "`pcx`" (PC paintbrush), and "`png`" (portable network graphics). MATLAB can use the built-in function `imread` to import the digital image into the computer system as shown in Listing 1.1.1. The imported image is stored in the $M \times N$ array `f=[f(m,n)]`, with line index $1 \leq m \leq M$ representing the vertical position, and column index $1 \leq n \leq N$ representing the horizontal position of each pixel within the image array, where $m, n \in \mathbb{Z}^+$. Please note that all the matrices and vectors in MATLAB are indexed from 1 and onward, while our analytical derivations will manipulate signal vectors and matrices from 0 and onward. These unmatched indices must be carefully handled when coding image processing algorithms using MATLAB. The image `f` can be converted from color to grayscale by MATLAB function `rgb2gray` as shown in the following MATLAB listing.

Listing 1.3.1: Grayscale digital image.

```
>> f = rgb2gray(f);
>> whos f
   Name        Size          Bytes       Class
   f           512x512     262144      uint8
```

Note that the data type of `f` in this example is `uint8` (*8-bit unsigned integer*). The number 8 specifies the number of binary bits required to store the data at a given quantization level is known as the *bit resolution*. This works hand in hand with radiometric resolution to specify the clarity of the captured image to be observed by human. For instance, a binary image has two pixel values (black or white) only, and has 1-bit resolution. A grayscale image commonly has 256 different gray levels ranging from black to white, and therefore has an 8-bit resolution, which means that each pixel (each entry in the array `f(m,n)` in MATLAB) takes value in one of the integer between 0 and $2^8 - 1$ (8-bit encoding). Within these 256 levels, 0 is black and 255 is white in the convention of MATLAB. The bit resolution of a digital image is also referred to as the *dynamic range* of an image. The quantization effect of the 1D signal will have the same effect in the 2D digital image. As a result, digital images with low dynamic ranges will look blocky.

The image array `f` with data type `uint8` can be printed or displayed by electronic devices, such as printer, and computer screen. Listing 1.3.2 applies the MATLAB built-in function `imshow` to display `f` on computer monitor. The MATLAB command `imwrite` can be applied to store the image in the PC file system in a selected image format. The example in Listing 1.3.3 saves the `uint8` image array `f` in `png` format with filename `'image.png'`.

Listing 1.3.2: Display digital image.

```
>> imshow(f);
```

Listing 1.3.3: Saving a digital image `f` to a `png` file with filename "`image.png`".

```
>> imwrite(f,'image.png','png');
```

The readers may have noticed that we have placed special attention to the data type of the image array `f`. This is because the `uint8` data type can take value in one of the integers within the range of 0 to 255. In general image processing applications, different mathematical computations will be applied to the pixel values, resulting in intermediate floating point or negative values, which cannot be fully represented by the `uint8` data and results in overflow, or known as `data type error` in MATLAB. As a result, it is a common practice to convert the `uint8` data into `double` to render the `data type error` problem for any computation in MATLAB. `double` refers to a double-precision data which uses 64-bit to represent the data.

There are a number of built-in data conversion functions in MATLAB. `typecast` is the general data conversion function which supports lossless conversion between different data types. However, it is more common to use the typecasting functions `double` and `uint8` for direct conversion. It should be noted that the data in `double` data type has 64-bit data, and it may contain floating point information, the direct conversion of a `double` data to a `uint8` data will truncate and round off the data to the nearest integer in the range of [0, 255]. The truncation is system-dependent such that the result of direct conversion may be different from system to system. However, the difference is not noticeable in most of the cases. Listings 1.3.4 and 1.3.5 show the usage of the conversion functions `uint8` and `double`, respectively.

Listing 1.3.4: Data type conversion from `uint8` to `double`.

```
>> f = double(f);
```

Listing 1.3.5: Data type conversion from `double` to `uint8`.

```
>> f = uint8(f);
```

All the images considered in the examples in this book are of size 512×512 unless otherwise specified. 512 is used because many operations and arithmetic with images can be simplified when the dimension (both row and column) of the image is of power of 2. However, the readers should understand that all operations discussed in this book are applicable to images with arbitrary size through some small and necessary modifications.

1.4 Current Pixel and Neighboring Pixels

After we have imported an image into MATLAB as an array, we can manipulate it as any other numeric vectors and matrices. Such manipulation is termed as *image processing*. To ease our discussions on the particular pixel under processing to other pixels in the same image, we define the term *current pixel* to describe this particular pixel. In the case of image denoising, the value of the *current pixel* is equal to the noiseless image pixel value plus noise. The actual noiseless pixel value will be estimated by the *image denoising algorithm*, using the current pixel and its surrounding pixels, also known as *neighboring pixels*. The collection of the *neighboring pixels* can be very different for each denoising algorithm. The collection of *neighboring pixels* are sometimes referred to as the *training window*. For example, a 3×5 rectangular window is drawn in Figure 1.5 which encloses the *neighboring pixels* of the *current pixel* (the pixel with a gray color background). Besides rectangular window, the shape of the window can be circular or other shape as desired. It is a common practice to form a training window to have odd numbers of rows and columns, such that the *current pixel* is located at the center of the window. In any special case where the window has an even number of rows or columns, or both, the location of the *current pixel* should be clearly specified to avoid ambiguity.

3 × 5 neighborhood

255	255	255	255	254	222	53	29	24	27
255	255	255	255	253	216	40	31	27	27
255	255	255	254	254	243	63	31	27	21
254	255	254	250	252	210	51	18	28	22
254	255	255	231	135	138	62	40	30	26
225	227	191	121	143	221	224	47	39	27
110	109	119	157	235	254	254	77	40	30
146	145	121	146	231	254	254	231	63	45
186	180	168	152	134	180	244	253	203	49
254	255	245	210	170	136	144	212	253	211

Current pixel

Figure 1.5 Current pixels and its neighborhood.

1.4.1 Boundary Extension

Images have finite physical sizes, when neighboring pixels are required in the algorithm for current pixel located at the boundary of the image, there will be a problem because of the nonexistence of some of the required neighboring pixels. To prevent this problem, *boundary extension* is employed. Boundary extension is a process to generate the pixels outside the boundary of the finite size image. In other words, boundary extension will help to generate the pixel $f[m,n]$, when $m < 0$, $m \geq M, n < 0$, and $n \geq N$ under some predefined rules. There are more than one image boundary extension methods, and the choice of the extension method would affect the final outcome of the processed image [32]. We shall adopt the *symmetric extension* method in this book, where an example of the extended image is shown in Figure 1.6(a). This is because symmetric extension preserves the pixel intensity continuity across the image boundary, and continuity is one of the major considerations in image denoising problems. It can be observed that the extension can be considered a titling up of multiple images in same size as that of the original image. Figure 1.6(b) shows a numerical example of such symmetric extension. This is also known as the *half pixel symmetric extension method*. Since most of the algorithm to be developed in this book has a processing kernel that is half pixel symmetric, the half pixel symmetric extension method is adopted in this book unless otherwise specified. The symmetric extended image g from f is given by

$$g = \begin{bmatrix} f_x & f_{ud} & f_x \\ f_{lr} & f & f_{lr} \\ f_x & f_{ud} & f_x \end{bmatrix}, \tag{1.7}$$

where f_{lr} is the horizontally flipped version of f (flipping f left to right), f_{ud} is the vertically flipped version of f (flipping f top to down), and f_x is the horizontally

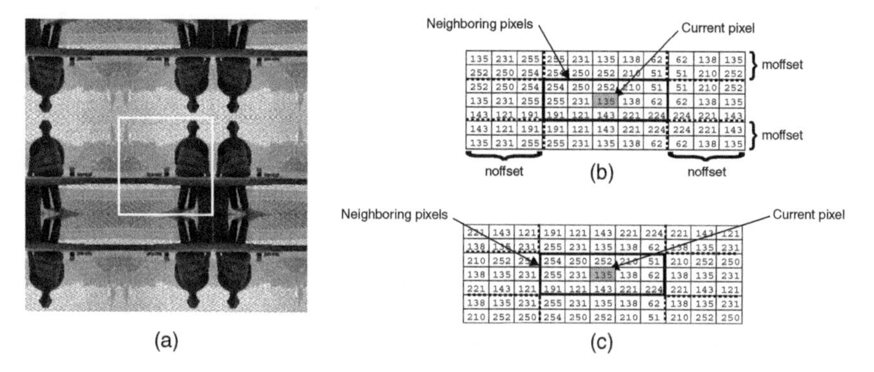

Figure 1.6 Boundary extension: (a) symmetric extension on image *Sculpture*, (b) pixels values of a 2×5 example image after half pixel symmetric extension with three pixels horizontally (`noffset=3`) and two pixels vertically (`moffset=2`), and (c) the same image in (b) after whole pixel symmetric extension of the same size.

flipped version of f_{ud} (flipping f_{ud} left to right). In most of the applications, only very limited size boundary extension is required, and therefore, we shall use the built-in function in MATLAB `padarray` to accomplish the extension operation. The size of the extension along the m and n axes are controlled by two parameters `moffset` and `noffset`, as shown in Figure 1.6(b), where `moffset=2` and `noffset=3`. Listing 1.4.1 shows the application of `padarray` realizing half pixel symmetric extension with `moffset=2` and `noffset=3` to the image `f` to obtain the extended image `g` as that showing in Figure 1.6(b). The parameters `'symmetric'` and `'both'` direct the padding to give mirror reflections of the input array `f` in the size of `[moffest,noffest]` before and after the first array element along each dimension.

Listing 1.4.1: Half pixel symmetric extension by MATLAB built-in function `padarray`.

```
>> moffset=2; noffset=3;
>> g = padarray(f, [moffset noffset],'symmetric','both');
```

Besides the half pixel symmetric extension method, there is also the *whole pixel symmetric extension method*, where the pixels on the boundary are *not* repeated in the extension, as shown in Figure 1.6(c). The MATLAB built-in function `wextend` extends real-valued input vector of matrix `f` according to the settings in the input parameters. The first and the second parameters are the extension span and the extension method, respectively, where `'2D'` specifies it is a two-dimensional extension and `'symw'` specifies it is a symmetric padding to replicate and mirror the boundary pixels, where `moffset=2` rows of pixels and `noffset=3` columns of pixels would be padded to each side of the input image f to realize an extended image g. The choice of the extension method will depend on the design of the filter to be applied in the denoising.

Listing 1.4.2: Whole pixel symmetric extension by MATLAB built-in function wextend.

```
>> moffset=2; noffset=3;
>> g = wextend('2D', 'symw', f, [moffset noffset]);
```

It should be noted that the boundary extension has enlarged the image by padding it with more pixels to provide necessary neighboring pixels to alleviate the lack of neighboring pixels when processing boundary pixels in the original image. For most of the image denoising results obtained with a size extended image. In this case, the denoised image has to be re-adjusted, usually by trimming, to restore to the same size as that of the original image.

1.5 Digital Image Noise

Noise is the variation of signal from its true value due to external or internal factors in image capturing, transmitting, and processing. Due to the quantic nature of light, the photon count at each pixel sensor is a stochastic process. Technically, some amount of noise will always be in every photo. There is nothing you can do to prevent this; it is a physical property of light and photography. The type and energy of the image noise naturally depend on the way the images have been acquired or transmitted. It follows that all image noise is a grainy veil in digital imaging, obscuring details and making the digital image appear significantly worse. In some cases, the digital images can be so noisy that it does not contain any useful information to the observers.

This is the reason why, when you take a photo with the lens cap on, the resulting digital image will not be totally black. Figure 1.7(a) shows the image taken by a DSLR with cap on. There will always be bright, and discolored pixels randomly scattered around the image. In this case, you can see the random pixels very easily by brightening the image through multiplying all pixel intensity with a constant 2, as shown in Figure 1.7(b). These random pixel intensities are the noise of the image. The problem becomes much worse in many consumer cameras and mobile phones due to their application of small sensor size and insufficient exposure. The issue of noise is so important that it is used as a valuable metric to determine the quality of the digital image sensor which helps to determine how well the camera will perform.

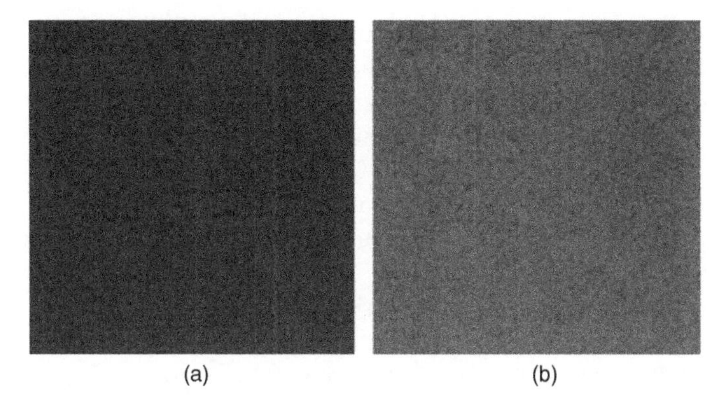

Figure 1.7 (a) A total dark image taken by a DSLR and (b) the same photo with all pixel intensity multiplied by 2.

The value $f[m,n]$ observed by a sensor at each pixel located at $[m,n]$ is a Poisson random variable whose mean $\upsilon[m,n]$ would be the ideal image. The difference between the observed image and the ideal image $f[m,n]-\upsilon[m,n]=\eta[m,n]$ is called the noise image, or simply the noise. The standard deviation of the Poisson variable $f[m,n]$ is equal to the square root of the number of incoming photons $\upsilon[m,n]$ received by the sensing element at $[m,n]$ during the exposure time together with the noise signal $\eta[m,n]$ that is the sum of a thermal noise and electronic noise. Therefore,

$$f=\upsilon+\eta. \tag{1.8}$$

The thermal noise is approximately additive and white, while the electronic noise is impulsive, which can be modeled by salt and pepper noise (SAP). At some level, we are all quite familiar with the concept of noise, yet it is crucial to understand it properly if we want to suppress it in order to maximize the quality of the digital image. Furthermore, we shall present the generic noise generation methods for various noise properties instead of using the MATLAB image processing toolbox built-in function `imnoise` such that we hope our reader will be able to generate the MATLAB functions for their own type of application specific noise in concern.

1.5.1 Random Noise

The random noise, also known as digital noise or electronic noise, is caused by own camera sensor and internal electronics, which introduce imperfections to an image. Noise is generated in multiple scenarios in the camera. The first scenario is when the image sensor gets heated up during operation, where the heat is high enough to stimulate electrons which will affect the electrical signal output of the sensor and thus generating the "thermal noise." These superfluous electrons will be mixed with the electric signal generated by the photo sensors with respect to the image that the camera is capturing. These electronic signals will be converted to an analog signal before leaving the image sensor. As a result, it is vivid that the analog signal of the captured image is contaminated before it even gets processed.

Theoretically, each photosite within the image sensor is independent. However, in practice the superfluous electrons generated in one photosite might also affect other photosites when they are in close proximity. This effect is especially vivid when the image sensor is small, and a lot of photosites are packed into a smaller area. In that case, the thermal noise within the digital image will not be uncorrelated from pixel to pixel, but it will be correlated with the corrupted noise.

Another common cause of noise is shooting at higher ISO settings.[2] As these settings basically magnify the light signal, they also magnify other unwanted signals such as background interference (e.g. heat sources). When we are photographing an area of low light, the background signals can be strong enough to compete with the signals from the limited light of the area that the camera is shooting.

1.5.2 Gaussian Noise

Gaussian noise is an idealized form of white noise. It is caused by random fluctuations in the signal. It is normally distributed. It can be generated in MATLAB by the following listing.

Listing 1.5.1: Additive Gaussian noise.

```
>> noise = sigma.*(randn([size(f)]));
```

where the generated noise power is $\sigma_\eta^2 = \texttt{sigma}^2$. Because of its mathematical tractability in both the spatial and the frequency domains, Gaussian noise models are used frequently in practice. Consequently, throughout this book, the focus will be on restoring digital images that have been corrupted by an additive white Gaussian noise (AWGN). More specifically, the noise, η, is assumed to be a wide-sense stationary (WSS) AWGN process with zero mean and constant variance σ_η, which is formed independent of the original noise-free image. The noise-corrupted image f is thus given by

$$f = \upsilon + \eta, \tag{1.9}$$

2 ISO, which stands for International Standards Organization, is the sensitivity to light as it pertains to either film or a digital sensor.

where υ and η are statistically independent. In the pixel domain, one has

$$f[m,n] = \upsilon[m,n] + \eta[m,n], \qquad m = 1,2,\ldots,M, \text{ and } n = 1,2,\ldots,N, \tag{1.10}$$

where $\eta[m,n]$ are independent and identically distributed (IID) Gaussian random samples with zero mean and variance σ_η^2. Analytically, the AWGN can be written with the normal distribution function $\mathcal{N}(\cdot,\cdot)$ as

$$\eta[m,n] \sim \mathcal{N}(0,\sigma_\eta^2), \quad \text{for } m = 1,2,\ldots,M, \text{ and } n = 1,2,\ldots,N. \tag{1.11}$$

The *Sculpture* image corrupted by Gaussian noise with $\sigma_\eta = 10$ and $\sigma_\eta = 50$ are shown in Figure 1.8(a) and (b), respectively. It can be observed that Figure 1.8(b) is visually more noisy than that of Figure 1.8(a) due to a greater noise variance (i.e. greater Gaussian noise power), and it is also reflected in their peak signal-to-noise ratio (PSNR) values (PSNR will be discussed in Section 1.8.2), where Figure 1.8(a) has a higher PSNR at 28.2 dB while that of Figure 1.8(b) is 14.2 dB. Compared with the noisy image with a Gaussian noise with $\sigma_\eta = 10$ added to a uniform tone image with the intensity of 128 shown in Figure 1.8(c), it is vivid that the noise-corrupted image is observed to be more noisy than the Gaussian noise-corrupted *Sculpture* image with the same noise power in Figure 1.8(a). The robustness of Figure 1.8(a) toward noise corruption is known as noise masking, where more details will be discussed in Section 1.9.

The noise to be added to the image can be specified by either the noise power, or the signal-to-noise ratio (SNR) of the corrupted image, where the first one is independent of the image, while the noise in the second one is normalized by the signal power of the noise-free image. To achieve a noise-corrupted image with a particular SNR, the SNR in decibels should be converted to linear scale first, which can be achieved by the following MATLAB listing.

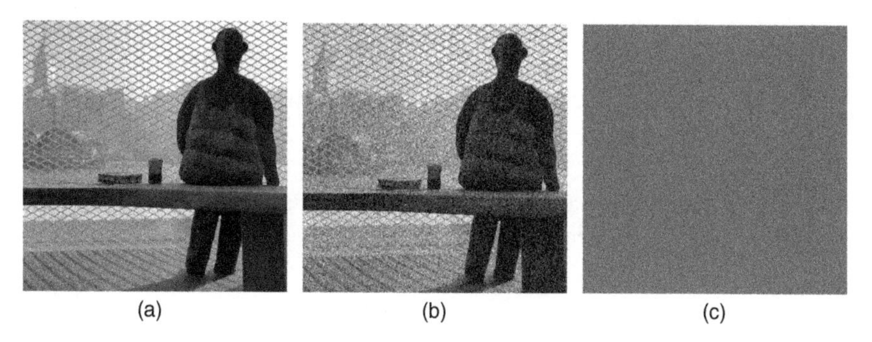

Figure 1.8 Additive Gaussian noise-corrupted (a) *Sculpture* image with zero mean and $\sigma_\eta = 10$ (PSNR=28.2 dB), (b) *Sculpture* image with zero mean and $\sigma_\eta = 50$ (PSNR=14.2 dB); and (c) uniform tone image with intensity level at 128 and $\sigma_\eta = 10$.

Listing 1.5.2: dB scale to linear scale.

```
>> snr = 10^(SNRdB/10);
```

At the same time, the power of the input image and noise image should be determined. The following MATLAB listing shows you how to determine the input image and noise image power, and then scale the noise image appropriately such that when `noise` is applied to the image `f`, the SNR of the resulting image will equal to `snr` in linear scale.

Listing 1.5.3: AWGN corrupted image with predetermined SNR.

```
>> f = im2double(f);                % convert f to double in [0,1] range
>> fpower = sum(sum(f.^2))));       % Power in f
>> noise = randn(size(f));          % zero mean Gaussian noise
>> npower = sum(sum(noise.^2)))); % noise power
>> rpower = (fpower/npower);        % signal-to-noise power ratio
>> noise = sqrt(rpower/snr)*(noise);  % reshape to specific SNR
>> f = f+noise;                     % add noise to f
>> g = imgtrim(g);                  % Truncating pixel values into [0,255]
```

The noise image is generated using MATLAB built-in random number-generating functions `randn`. Noted that the AWGN corrupted image generated by adding the noise image to the natural image might result in pixel intensity overflow (pixel intensity greater than 255) or underflow (pixel intensity less than 0). Therefore, the function `imgtrim`, as shown in Listing 1.5.4, is applied to truncate all negative pixel values to 0 and capping all pixel values greater than 255 to 255.

Listing 1.5.4: Pixel values truncation.

```
function g=imgtrim(f)
  g = f;
  g(g>255)=255;
  g(g<0)=0;
end
```

Noted that `imgtrim` is a nonlinear process. The difference in SNR of the two images f and g will be untractable.

1.5.2.1 Noise Power Estimation

At the beginning of Section 1.5.2, the concept of SNR has been introduced; however, the noise level σ_η^2 is unknown in practice. One method to estimate the noise variance of the AWGN corrupted image is based on the assumption that an image has many regions of almost uniform intensity and that most changes in these regions of insignificant variations are due to the noise. This assumption is generally valid for many real-world images. The background of a scene is an example of such a region of insignificant variations. Also, the noise η is assumed to have a constant variance σ_η^2 throughout the image. This is a direct consequence of the fact that the noise is assumed to be a WSS process. The local variance estimates of all window masks of size $\mathbb{W} \times \mathbb{W}$ pixels, centered at every pixel of the image, are then calculated. The choice of the window size over which to estimate the local variance is important. The following MATLAB listing implements the discussed noise variance estimation using local statistics.

Listing 1.5.5: Noise estimation by local statistics.

```
function sigma = nestls(f,w)
halfw=floor(w/2);
index=1;
[m n] = size(f);
wsq = w*w;
sigma = zeros(1,(m-w+1)*(n-w+1));
for x = halfw+1:m-halfw
    for y = halfw+1:n-halfw
        localwin = f(x-halfw:x+halfw,y-halfw:y+halfw);
        localrow = reshape(localwin,[1,wsq]);
        sigma(index) = sqrt(var(localrow));
        index=index+1;
    end
end
end
```

The function `nestls` will return an array that contains the variance of each $\mathbb{W} \times \mathbb{W}$ window. The noise variance of the image can be obtained by examining the distribution of the local variance, where the mode (i.e. the most frequent value) of the local variance distribution (histogram) was shown to be a reasonable estimator of the noise variance [26]. To obtain an accurate local statistics estimation, the window size has to be at least 5×5, but it should also be small enough to ensure local signal is stationary. Empirical result shows that both 5×5 and 7×7 windows work well for most natural image. Figure 1.9(a) illustrates the histogram of these local variance estimates obtained by the MATLAB Listing 1.5.6 with a uniform tone (with an intensity of 128) image corrupted by AWGN with $\sigma_\eta = 50$ as input and the mask size is 7×7. It can be observed that the mode of the histogram (i.e. the standard deviation value where the peak of the histogram is found) is at 49.7119, which is very close to the true noise standard deviation $\sigma_\eta = 50$.

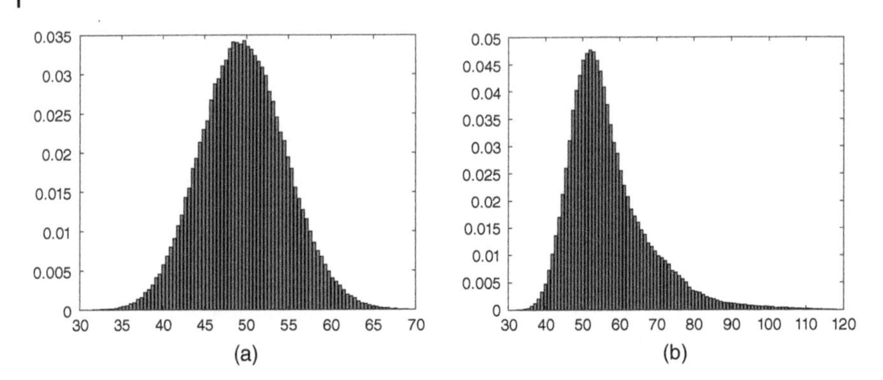

Figure 1.9 The histogram of the noise variance computed from 7×7 masks of (a) uniform tone (128) image corrupted by AWGN with $\sigma_\eta = 50$; (b) *Sculpture* image corrupted by AWGN with $\sigma_\eta = 50$.

Listing 1.5.6: Histogram of noise variance.

```
>> sigma = nestls(f,7);
>> nbins = 101;
>> [h,t] = hist(sigma,nbins); h=h/sum(h);
>> bar(t,h);

>> [max_val,max_idx] = max(h);
>> mode_val = t(max_idx);
```

When the same method is applied to the *Sculpture* image corrupted by AWGN with $\sigma_\eta = 50$, the histogram of the locally estimated σ_η with a 7×7 window is plotted in Figure 1.9(b). It can be observed that the mode of the histogram is at 52.0468 which is very close to the true noise standard deviation $\sigma_\eta = 50$. Although the difference between the mode of the histograms of the two estimations are very small, the distribution of the estimations are very different due to the local structures of the two images.

1.5.2.2 Noise Power Estimation Base on Derivative

As aforementioned in Section 1.5.2.1 that the texture *Sculpture* image is one of the reasons that affects the noise power estimation accuracy obtained by the windowed local statistics method in previous section. A simple method to take the texture into consideration is through the normalized 2D derivative. A simple first-order derivative along the horizontal and vertical directions can be obtained through

$$\left|\frac{\partial f[m,n]}{\partial m}\right| = \frac{f[m+1,n] - f[m,n]}{\sqrt{2}}, \tag{1.12}$$

$$\left|\frac{\partial f[m,n]}{\partial n}\right| = \frac{f[m,n+1]-f[m,n]}{\sqrt{2}}, \tag{1.13}$$

which measures the rate of pixel intensity change between adjacent pixels. The 2D derivative image can be obtained through the following MATLAB listing.

Listing 1.5.7: 2D derivative.

```
function D = deriv(f)    % first order 2D derivative on f
  D = double(f);
  [m n] = size(D);
  D = (D(1:m-1,:) - D(2:m,:))'/sqrt(2);
  D = (D(1:n-1,:) - D(2:n,:))'/sqrt(2);
end
```

The above derivative function `deriv` measures the pixel intensity variation within a 2×2 window. When the image is assumed to be homogenous, this 2D derivative will be equivalent to the noise added to the image. Furthermore, the noise is Gaussian distributed; therefore, the variance σ_η can be estimated through median of absolute difference (MAD) estimator of the derivative of the image `f` as

$$MAD(deriv(f)) = MAD(\eta) = \sigma_\eta \sqrt{2}\mathrm{erf}^{-1}(0.5) \approx 0.6745\sigma_\eta, \tag{1.14}$$

where `erf` is the error function [43]. Other research works that make use of the quartile method on MAD for noise power estimation will also result in a similar constant as that depicted in Equation 1.14 with the constant 0.6745 replaced by

$$0.6745 \approx Quantile\left(\eta, \frac{3}{4}\right). \tag{1.15}$$

The quantile function $Quantile(x, y)$ refers to cut points dividing the range of a probability distribution of x into continuous intervals with equal probabilities. The quantity $y = \frac{3}{4}$ in Equation 1.15 refers to the computation of the third quartile of the noise image $x = \eta$. If η is AWGN with zero means, then the difference between the two sides of the equation shown to be less than 1.025×10^{-5} which is 0.0015%. As a result, the noise σ_η can be implemented with the following MATLAB listing.

Listing 1.5.8: Noise estimate via derivative.

```
function sigma=derisigmaest(f) % derivative based sigma estimate
  D = deriv(f);
  sigma = mad(D(:),1)/0.6745;
end
```

The `mad` estimator will obtain an estimated $\sigma_\eta = 51.6312$ for the *Sculpture* image corrupted by additive Gaussian noise with $\sigma_\eta = 50$. It can be observed that the estimated σ_η is close to the real σ_η. More robust noise estimation methods will be discussed in Chapters 3 and 4.

1.5.3 Salt and Pepper Noise

The origin of *salt and pepper noise* (SAP) is the photon noise, caused by the randomness of the photons in the scene being photographed. Light emits and reflects off everything that can be seen, but it does not happen in a fixed pattern, which will result in graininess in the photographed images. For example, a very dim light bulb may emit an average of 1000 photons per second, but the number of photons emitted in each individual second will vary, say in one second it may emit 986 photons, and 1028 photons in the following second, and so on. It is the average number of photons emitted by the light bulb describes the illumination power of the light bulb, and thus we usually do not care about the details of how many photons being emitted in each second. Since each area within the photo will suffer from this kind of randomness in the number of photons which will translate to the brightness of that particular image area. Such image brightness variation is what photographers call *shot noise* in an image. The shot noise reserves the discrete and random nature from the photon.

To be precise, this section discusses SAP, also known as impulse noise or binary noise. This noise will degrade the input by causing a sharp and sudden disturbances to selected image pixels. The result will be randomly scattered white or black pixels over the image with the following probability

$$p(f[m,n]) = \begin{cases} \gamma, & \text{with } f[m,n] = f[m,n], \\ \beta, & \text{with } f[m,n] = a, \\ \alpha, & \text{with } f[m,n] = b, \end{cases} \tag{1.16}$$

where $p(\cdot)$ is the probability measure on the occurrence of the operand. The sum of probability $\alpha + \beta + \gamma = 1$. Usually, $a = 0$ and $b = 255$, which cause the corrupted pixels to be bright white (the salt) or dark black (the pepper), and hence the name salt and pepper noise. MATLAB has a built-in SAP image corruption function `imnoise(f,'salt&pepper',nd)`, where the last parameter `nd` is the noise density which equals $\alpha + \beta =$ `nd` in Equation 1.16, and $\alpha = \beta$. In this book, we shall provide our own implementation as shown in `sapnoise` with $a = 0$ and $b = 255$ for a `uint8` input image `f`.

Listing 1.5.9: Salt and pepper noise (SAP).

```
function g = sapnoise(f,nd)   % salt and pepper with density nd/2
  uloc = rand(size(f));       % uniform distribution between 0 and 1
  g = f;                      % pepper noise for uniform random number
  g(uloc<nd/2) = 0;            % < half of the specified noise density
  g(*uloc>=nd/2)&(uloc<nd)) = 255;  % salt noise
end
```

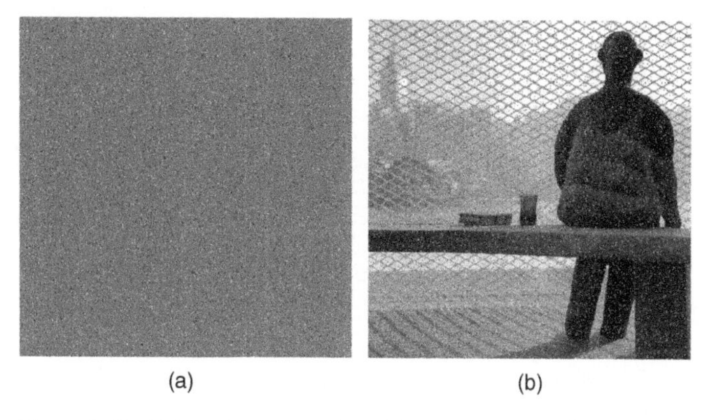

Figure 1.10 Salt and pepper noise with total noise density being 0.05 over (a) image with gray level 128; and (b) *Sculpture* image (*PSNR* = 17.7 dB).

The function that generates an array contains a uniform random variable that spans the array locations of the image `f`. We select all the array locations that have values within [0,`nd`/2] as the pixel locations in the image f to be corrupted with pepper noise. While those array locations having values in the range [`nd`/2,`nd`] as the pixel locations in the image f to be corrupted with salt noise. The resulting SAP corrupted image is shown in Figure 1.10 with noise density being `nd` = 0.05.

1.6 Mixed Noise

Natural images can be corrupted by both AWGN and SAP at the same time. The power of these two kinds of noises is independent although in reality, they are not. Let us consider Listing 1.6.1 that corrupts the grayscale *Sculpture* image by AWGN with $\sigma_\eta = 50$ first and then by SAP with total noise density of 0.05.

Listing 1.6.1: Mixed noise – AWGN then salt and pepper noise.

```
>> noise = 50.*(randn([size(f)]));
>> f = double(f);
>> f = f+noise;
>> f = imgtrim(f);
>> g = sapnoise(f,0.05);
>> g = imgtrim(g);
```

The noise-corrupted image `g` is shown in Figure 1.11(a). When the order of noise corruption is reversed, such that the SAP is first applied, followed by AWGN with the same noise variance, as shown in Listing 1.6.2. The resulting noise-corrupted image `g` is shown in Figure 1.11(b).

Listing 1.6.2: Mixed noise – salt and pepper noise then AWGN.

```
>> noise = 50.*(randn([size(f)]));
>> f = double(f);
>> f = sapnoise(f,0.05);
>> g = f+noise;
>> g = imgtrim(g);
```

The image in Figure 1.11(b) is observed to be less corrupted with SAP when compared to Figure 1.11(a). Furthermore, the objective PSNR of the noisy image in Figure 1.11(a) is 13.4351 dB which is lower than the 13.7241 dB obtained from Figure 1.11(b) which is consistent with that of the subjective results. This example shows that the noise corruption process is a nonlinear process, and the resulting image is process-dependent. There is no definite procedure to model images corrupted with mixed noise. However, to allow the reproducibility of the results presented in this book, we shall use the mixed noise-corrupted image in Figure 1.11(a) throughout the whole book.

The nonlinear property of the noise corruption process can also be observed from the noise power of the corrupted images. Although the AWGN and SAP are independent processes, their power is not additive to the corrupted image. This is because it is highly likely that some of the pixels will be corrupted by both AWGN and SAP. When that happens, the pixel value will still be either 0 or 255 after `imgtrim`, which is the same as applying SAP alone. In other words, the total noise power corrupting the image is not the sum of the noise power of individual noise sources. Therefore, the PSNR of all the images are not the same even though the individual noise power that have been applied to the images are the same.

(a)

(b)

Figure 1.11 *Sculpture* image corrupted with (a) AWGN with $\sigma_\eta = 50$ and then SAP with density 0.05 ($PSNR = 13.4351$ dB); (b) SAP with density 0.05 and then AWGN with $\sigma_\eta = 50$ ($PSNR = 13.7241$ dB).

Lastly, it should be noted that the noise is generated by random process. In other words, the noise-corrupted images obtained by the MATLAB listing will be different every time the MATLAB listing is executed. As a result, in order to provide a consistent discussion, and fair comparison, researchers usually will execute the MATLAB listing once and then save the noisy images, which will be recalled and used to test the performance of the denoise algorithm. In the case of this book, the noisy image being applied to all chapters is stored in the companion website of this book, for which the readers can load into MATLAB and test the algorithm presented in this book, where we expect the reader should be able to obtain the same quality measures as those presented in the book.

1.7 Performance Evaluation

Human eye interprets the information in an image by classifying the image into different feature zones and determining the image quality by looking for visible artifacts, which refers to the features that should not exist in the particular feature zones. A rough classification of the different feature zones of a natural image will consist of

1. *Homogeneous*: The variations of the grayscales within these zones are small (or smaller than a predefined quantity), which makes noise (large pixel value variations) in these regions to be easily noticeable.
2. *Textured*: Regions with repetitive patterns and structures at various scales and orientations. Human eyes are not very sensitive to pixel value variations within these zones, and noise is difficult to be noticed in these regions.
3. *Edges*: The edges separate two homogeneous regions with different mean grayscales, which makes the noise in this zone readily noticeable.

Subjective quality assessment assesses the image quality through human eye, which is often considered to be the only *correct* way to evaluate the image quality. The subjective *mean opinion score* (MOS) is a popular method to achieve statistical significance. To achieve a satisfactory assessment results, which can be considered to be general and reliable, the number of participants should be as large as possible. As a result, each assessment will require an adequate number or participants and a series of tests are required, making the experiment extremely time-consuming and expensive. Therefore, a great deal of efforts have been made in recent years to develop objective image quality metrics that correlate with perceived quality measurement, which are the topics in Section 1.8.

The performance assessment of image denoising algorithms can be categorized into objective and subjective assessments, and they are just the two faces of the mirror. Since the denoised images are to be perceived by human eyes, subjective analysis is considered to be the final quality assessment of the denoised image. However, one's medicine is the other's poison. It is difficult if not impossible to provide a subjective analysis of the denoised image as it requires time and money and is highly inconvenient. Not to mention that there is no commonly accepted subject quality measure or feature sets for all varieties of image denoising problems. Researchers are devoting massive efforts to develop different objective quality assessment algorithms that take the HVS into consideration (to model and approximate the behavior of human vision) such as to provide an objective mean to compare the visible artifacts generated throughout the denoising process. These algorithms give objective quality score that mimics the subjective quality measure for the image under test, without going through the subjective quality analysis. The objective scores (which are sometimes referred to as index) of different quality assessment algorithms depend on how the visible artifacts are quantified and also the sources of the reference data for comparison. Therefore, it is important for the readers to understand the definition of visible artifacts in terms of their appearances; and also the sources of the reference data, such that they can make appropriate choices of the objective quality assessment methods to be applied for their own purposes. In this section, we shall first introduce different classification of quality assessment algorithms according to the sources of the reference images.

The source of the reference image adopted in different quality assessment algorithms categorizes the algorithms into three groups, including *full reference image quality index* (FRIQ), *no reference image quality index* (NRIQ), and *reduced reference image quality index* (RRIQ). Among various image quality indices, the interest of this book is the FRIQ because we have no difficulties to obtain the reference image in our analysis. The FRIQ scores the quality of the denoised image by comparing it with a reference image, which is also known as the undistorted image. Different measures use different parameters of the image to estimate the quality score of the denoised image with reference to the undistorted image. A list of commonly applied FRIQ measures together with their analytic backgrounds will be discussed in Section 1.8, in which all the algorithms focus on some kinds of measures of the absolute difference in pixel intensities between the denoised image and the undistorted image.

In Section 1.9, we shall further our discussions of a benchmark FRIQ that considers the HSV, which is known as *structural similarity* (SSIM) index [50]. The SSIM takes an in-depth look on the impact of the image structure on the assessment of the image quality. The readers should note that neither subjective nor objective assessments could be used alone. It is always more convincing when both quality measures are applied together, or at least a limited subjective quality measure is applied to assist the objective assessment of the denoised image quality.

1.8 Image Quality Measure

The task of performing noise reduction (also known as *denoising*) is synonymous with improvement in image quality. In general, most of the performance evaluations for these various noise reduction methods are done on small images contaminated with artificial noise (Gaussian, Poisson, salt and pepper, etc.), which is artificially added to a clean image to obtain a noisy version. Measuring the performance of a noise reduction algorithm on images corrupted by artificial noise will give an accurate enough picture of the denoising performance of the algorithm on real digital cameras or mobile images.

In this book, we shall only focus on the objective quality metric $Q(g, \upsilon)$ that correlates the perceived difference (quality) between the denoised image g and the noise-free reference image υ, which also satisfies the following conditions:

1. *Symmetric*: $Q(g, \upsilon) = Q(\upsilon, g)$.
2. *Boundedness*: $Q(g, \upsilon) \leq B$ for a constant B.
3. *Unique maximum*: $Q(g, \upsilon) = B$ if and only if $g = \upsilon$ (no distortion between the two images under concern).

Among various quality metrics, the *mean squares error* (MSE) and the *PSNR* are two commonly used metrics. These metrics are convenient in their simplicity to compute. Their physical meanings are similarity measurement by comparing the intensity of the two images in a pixel-by-pixel fashion, where neither the structure of the image nor human perception to the image features are considered. Therefore, they may not match well with the subjective quality and may lead to undesirable results in some cases.

To improve the assessment accuracy, the similarity measurement has to be modified to make it compatible with the HVS. The *texture peak signal-to-noise ratio* (tPSNR) and the *flat peak signal-to-noise ratio* (fPSNR) are modified from the basic PSNR by considering the image context, such that it will only compute the PSNR with respect to the pixels that fall into the chosen context. The two contexts are the texture (hence the synonym "t") and the smooth (also known as flat and hence the synonym "f") regions of the image because humans have very different tolerances for noise in these two different image areas. The tPSNR and fPSNR in Section 1.8.3 can be considered as our first step to perform objective similarity measure in response to the HVS. A more sophisticated and widely applied HVS-modified similarity measure, the SSIM [50] will be presented in Section 1.9.

Once an objective quality metric is chosen, it can be applied to evaluate the performance of an image denoising algorithm through a scheme, as shown in Figure 1.12. This scheme considers a noise-free reference image (the undistorted image) and a noisy image embedded with noise on known properties. Denoising

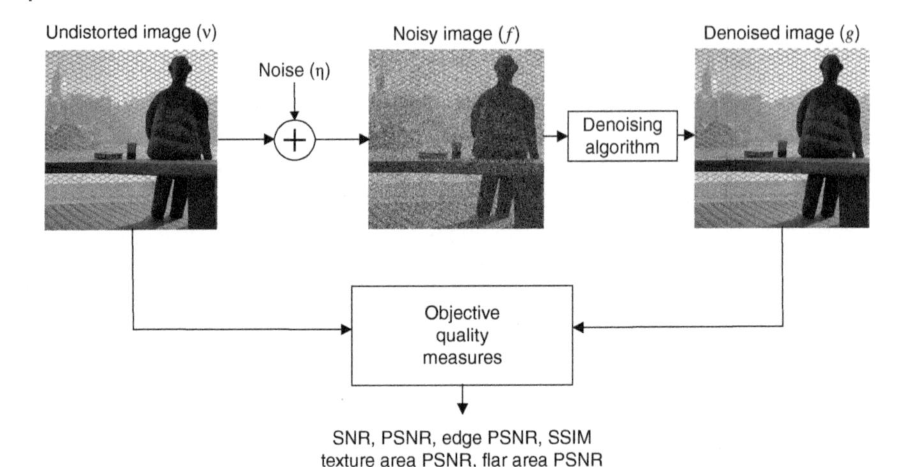

Figure 1.12 Image denoising quality computation.

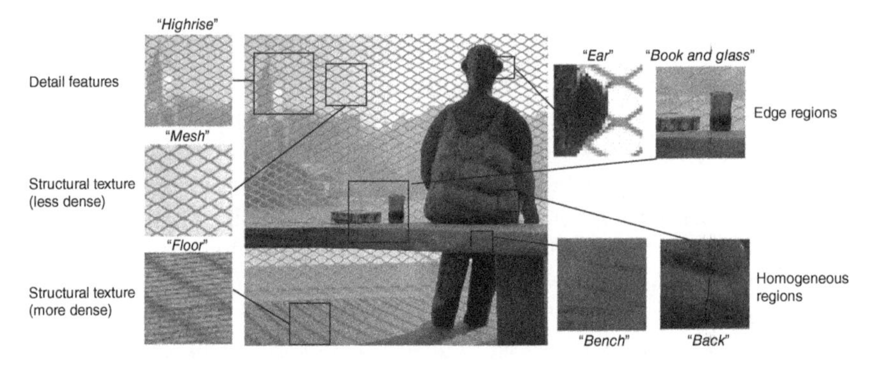

Figure 1.13 Different regions of interest in *Sculpture* image.

is performed on the noise image to obtain the denoised image. The noise free reference image and the denoised image will be applied to the chosen metric which will compute a numeric value that can be used to rank the denoising performance objectively. The performance of the image denoising algorithm should also be evaluated in a subjective manner. The *Sculpture* image as shown in Figure 1.13, will be applied in this book for both objective and subjective performance application. The *Sculpture* image contains complex features of texture regions, edge regions, and homogeneous regions. To simplify our discussions in this book, we have named several selected regions that are dominated by different image features, as shown in Figure 1.13. The *detail features* outlining the *highrise* blended behind the mesh at the backdrop. This region will be applicable to evaluate the performance of denoising algorithm in preserving image details subjected to texture masking problem. The *less dense texture* region is dominated by the mesh which can assist us to evaluate the performance of the denoising algorithm in restoring regular and structural texture in the image. The *more dense texture* region captured the shades of the mesh projected onto the ground, which will help us to evaluate the performance of the denoising algorithm in maintaining the weak texture structure in the image. To evaluate the performance of the image denoising algorithm in edge regions, we shall focus on the image objects with abrupt intensity changes in outlines, such as the *ear* of the sculpture, and the *book and glass* on the bench. The homogeneous regions, such as the smooth *bench* sidewall and the *back* of the sculpture with large granite-like patterns, will be considered when investigating the noise artifacts introduced into the denoised image by the image denoising algorithm.

1.8.1 Mean Squares Error

Intuitively, the noise-corrupted image can be regarded as the sum of the noise-free reference image and an error signal (also known as error image). Furthermore, if denoising is considered to be the signal separation process that separates the image from noise with respect to the image obtained in this separation process, a measurement on how good the separation is performed can be obtained by measuring the mean squares differences between the separated image and the original noise-free image. Such a measure is known as the *mean squares error* (MSE). Starting with the computation of the error image e (also known as the difference image) between the denoised image array g and the noise-free reference image array υ both of size $M \times N$ by

$$e[m,n] = \upsilon[m,n] - g[m,n]. \tag{1.17}$$

The MATLAB source code in Listing 1.8.1 implements the function computing the error image.

Listing 1.8.1: Error image.

```
function e=imageerr(g,v)
  e = (double(v)-double(g));
end
```

With the availability of the error image, the total error between the two images is given by $\sum_{m=0}^{M-1}\sum_{n=0}^{N-1} e[m,n]$. However, the elements in $e[m,n]$ have both positive and negative values. Therefore, it is more reasonable to consider the magnitude of $e[m,n]$. The *mean absolute error* (MAE) (also known as mean absolute difference) provides such a quality factor between the denoised image array g and the noise-free reference image array v both of size $M \times N$.

$$MAE = \frac{1}{M \times N} \sum_{m=0}^{M-1} \sum_{n=0}^{N-1} |e[m, n]|. \quad (1.18)$$

The MATLAB source code in Listing 1.8.2 implements the function to compute the error image.

Listing 1.8.2: Mean absolute error (MAE).

```
function mae_value=mae(g,v)
  e = imageerr(g,v);
  mae_value = mean(mean(abs(e)));
end
```

In particular, the most popular quality factor is the MSE, which is equivalent to the computation of the squares power of the error signal $e[m, n]$. The MSE is defined as

$$MSE = \frac{1}{M \times N} \sum_{m=0}^{M-1} \sum_{n=0}^{N-1} (e[m, n])^2. \quad (1.19)$$

The MATLAB source code in Listing 1.8.3 implements the MSE function.

Listing 1.8.3: Mean squares error (MSE).

```
function mse_value=mse(g,v)
  e = imageerr(g,v);
  mse_value = mean(mean(e.^2));
end
```

It is also common to give the MSE, `mse_value`, through the square root operation to generate a value that resembles the meaning of average pixel error of the two images, which is known as the *root mean squares error* (RMSE).

$$RMSE = \sqrt{(MSE)}. \quad (1.20)$$

The MATLAB source code in Listing 1.8.4 computes the RMSE by means of the function `mse`.

Listing 1.8.4: Root mean squares error (RMSE).

```
function rmse_value=rmse(g,v)
  rmse_value = sqrt(mse(g,v));
end
```

It should be noted that g and v should have the same array size to avoid runtime error.

At this particular moment, we would like to sidetrack and discuss the Frobenius norm which can be obtained by root sum of squares error, and is given by $||e||_F$, where $||\cdot||$ is the standard norm operator and the subscript F stands for the Frobenius norm, given by

$$||e||_F = \sqrt{\sum_{m=0}^{M-1}\sum_{n=0}^{N-1}(e[m,n])^2}. \tag{1.21}$$

The Frobenius norm measures the *distance* between matrices g and f. Noted that the Frobenius norm is symmetric such $||g - f|| = ||f - g||$ which is similar to that of other distance functions.

1.8.2 Peak Signal-to-Noise Ratio

The MSE does not consider the dynamic range of the image but only the absolute error in between two images and is therefore biased. Such bias can be removed by normalization. The PSNR is the most commonly used normalized objective quality metric for denoised image quality assessment. The denominator of the PSNR is the MSE, while the numerator is the highest dynamic range achievable by the image function under consideration, which is also known as the ratio between the maximal power of the reference image and the noise power of the denoised image. It is represented in the logarithmic domain in decibels (dB) to take care of the wide dynamic range of the signal power. The PSNR of the n-bit grayscale image is computed as

$$PSNR = 10\log_{10}\left(\frac{(2^n-1)^2}{MSE}\right). \tag{1.22}$$

For example, an 8-bit/pixel grayscale image has $(2^n - 1)^2 = 255^2 = 65025$ levels being the numerator of its PSNR. Listing 1.8.5 will compute the PSNR with $n = 8$ in Equation 1.22 as the maximum value in the data range of the input image array.

Listing 1.8.5: Peak signal-to-noise ratio (PSNR).

```
function psnr_value=psnr(g,v)
  psnr_value = 10*log10((255^2)/mse(g,v));
end
```

The above-discussed quality metrics can be easily extended to color images by treating each color channel independently as a grayscale image. In the case of color images in RGB domain, the PSNR of the three color channels is first computed and then recombined to give the final PSNR by averaging as

$$PSNR_{RGB} = (PSNR_{red} + PSNR_{green} + PSNR_{blue})/3, \tag{1.23}$$

where $PSNR_{red}$, $PSNR_{green}$, and $PSNR_{blue}$ are the PSNR values for the red, green, and blue channels of the color image computed with Equation 1.22, respectively. Without loss of generality, the rest of the book will use PSNR to imply both the $PSNR$ in Equation 1.22 for grayscale images and $PSNR_{RGB}$ in Equation 1.23 for color images, depending on the context.

The PSNR is widely used because it is simple to calculate, has clear physical meanings, and is mathematically easy to deal with for optimization purposes. High PSNR value of the denoised image is more favorable because it implies less distortion. However, the PSNR measure is not ideal. The major shortcoming is that the signal strength is estimated by the highest dynamic range of the image that can possibly achieved, which is $2^n - 1$, rather than the actual signal strength of the image. Furthermore, PSNR does not consider the HVS. It has been widely criticized for not correlating well with subjective quality measurement. One of such quality is the preservation of edges in the denoised image. Otherwise, the PSNR is considered to be able to provide an acceptable measure for comparing the denoising results.

1.8.3 Texture and Flat PSNR

A critical shortcoming of the MSE and the PSNR is that they are not comparable with the HVS. This problem is vivid in the denoised images in Figure 1.14. In this example, three AWGN noise-corrupted *Sculpture* images are shown in Figure 1.14(a)–(c) created by Listing 1.8.6 (there are two particulars about this MATLAB source that the readers should be aware of: (i) there are functions `tmap` and `imgtrim` inside the code that will be discussed in Sections 1.8.4 and 1.10, respectively; (ii) there are some twists on the parameters to create images with the same PSNR because the noise depends on texture contents in the image, which is an unknown before we run the code). There is no argument that the image (c) has better visual quality, while (a) and (b) are both visually observed to be seriously degraded in certain way; however, the PSNR of these images is all close to 14.2 dB. Besides having almost the same PSNR, we also observe several interesting phenomena from these three images. First, the AWGN is targeted to corrupt the texture-rich regions of the *Sculpture* image to generate

Figure 1.14(b), while for Figure 1.14(c), the AWGN is selectively corrupting the homogeneous regions of the *Sculpture* image. It is vivid that the AWGN is highly visible in the homogeneous image regions, such as the *bench* and the *back* of the *Sculpture* image. In particular, Figure 1.14(c) has the *highrise* in the background being completely washed out. The wash out problem will be discussed again in Chapter 2, together with an exercise in Chapter 2 about the image artifact. On the other hand, the AWGN noise observed in the texture-rich regions is almost no different among the three figures in Figure 1.14, even though there is no AWGN in the texture-rich region of Figure 1.14(b), which shows that the AWGN has no significant impact on texture-rich regions. Second, spatially correlated noise in Figure 1.14(b) is obviously more annoying than that of the noisy image is Figure 1.14(a). The same is also true for Figure 1.14(c) when compared to that of Figure 1.14(a) around the object edges (such as the *ear* and the *book and glass* of the *Sculpture* image), where salt and pepper alike noise appearing along the object outline. Thus, under given noise variance, it is more important to suppress spatially correlated (give raise by the spatial selectivity) noise. Third, this example shows the inadequacy of standard quality metrics – all three images have the same PSNR in Figure 1.14, but obviously they have different visual qualities. This is because the HVS perceives pixels differently depending on their visual features, while the PSNR considers all pixels to be the same. As a result, although being an objective and simple measure, the PSNR might lead to a totally wrong quality measurement result with respect to human observer.

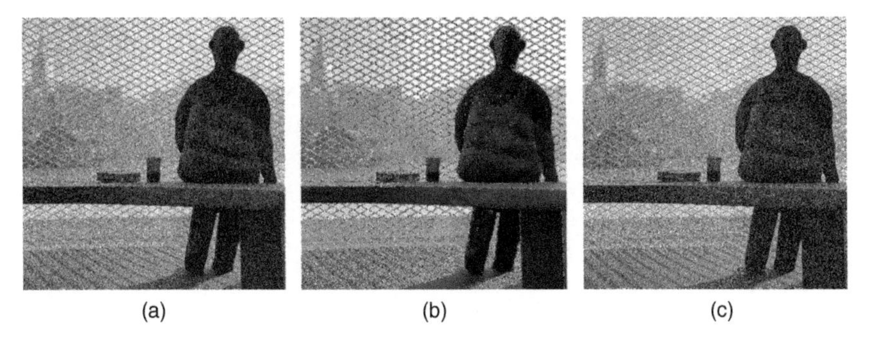

Figure 1.14 The *Sculpture* image corrupted by AWGN: (a) uniformly across the whole image with $\sigma_\eta = 50$; (b) in texture area alone with $\sigma_\eta = 2.59$; (c) in flat (homogeneous) area alone with $\sigma_\eta = 1.47$. Note that the PSNR of all three images is close to 14.2 dB.

Listing 1.8.6: Image corrupted with spatial correlated noise.

```
>> noise_gaussian_50=randn(size(sculpture))*50; % AWGN with sigma=50
>> tm=tmap(sculpture,100);                      % texture map
>> sculpture=double(sculpture);
>> snoise= sculpture+noise_gaussian_50;
>> snoie=imgtrim(snoise);        % noise image without texture preference
>> tnoise = noise_gaussian_50.*tm*2.59; % texture area correlated noise
>> ts = sculpture+tnoise;                % scales noise power
>> ts=imgtrim(ts);
>> fnoise = noise_gaussian_50.*(1-tm)*1.47;
>> fs = sculpture+fnoise;        % homogenous area correlated noise
>> fs=imgtrim(fs);
>> psnr(sculpture,snoise)
>> psnr(sculpture,ts)
>> psnr(sculpture,fs)
>> figure; imshow(uint8(snoise)); figure; imshow(uint8(ts)); figure;
    imshow(uint8(fs));
```

A simple step to improve the correlation between PSNR and visual quality of the denoised image is to incorporate the differentiation of pixels perceived by the HVS [22]. This can be achieved by decoupling the PSNR into two different pixel groups: the fPSNR and the tPSNR. These two new PSNRs will require us to define two new MSE measures

$$fMSE = \frac{\sum_{m=0}^{M-1} \sum_{n=0}^{N-1} (s[m,n]e[m,n])^2}{\sum_{m=0}^{M-1} \sum_{n=0}^{N-1} s[m,n]}, \tag{1.24}$$

$$tMSE = \frac{\sum_{m=0}^{M-1} \sum_{n=0}^{N-1} ((1-s[m,n])(e[m,n]))^2}{\sum_{m=0}^{M-1} \sum_{n=0}^{N-1} (1-s[m,n])}, \tag{1.25}$$

where $s[m,n]$ is the texture map for $v[m,n]$, with 1 assigned to the array location where the corresponding pixel is located in the smooth area, and 0 is assigned to pixel locations in the texture area. The two new PSNRs are therefore defined as

$$fPSNR = 10\log_{10}\left(\frac{(2^n-1)^2}{fMSE}\right), \tag{1.26}$$

$$tPSNR = 10\log_{10}\left(\frac{(2^n-1)^2}{tMSE}\right). \tag{1.27}$$

The MATLAB implementation of these two PSNR computations are given by Listings 1.8.7 and 1.8.8, where both functions make use of the texture classifier `tmap` to compute the $s[m,n]$. The texture classifier `tmap` will be discussed in Section 1.8.4.

Listing 1.8.7: tPSNR.

```
function tpsnr_value=tpsnr(g,v,th)
  e = imageerr(g,v);
  s = tmap(v,th);
  ee = ((e.*s).^2);
  tmse = sum(ee(:))/sum(s(:));
  tpsnr_value = 10*log10(255^2/tmse);
end
```

Listing 1.8.8: fPSNR.

```
function fpsnr_value=fpsnr(g,v,th)
  e = imageerr(g,v);
  s = 1-tmap(v,th);
  ee = ((e.*s).^2);
  fmse = sum(ee(:))/sum(s(:));
  fpsnr_value = 10*log10(255^2/fmse);
end
```

It is worth noticing that, indifferent with PSNR, both fPSNR and tPSNR are not symmetric functions, since the texture map needs to be computed on the original, noise-free image v. Furthermore, it should be noted that the MSE can be formed as a convex combination of the fMSE and tMSE. Therefore, it is vivid that the PSNR always lies between the fPSNR and the tPSNR.

1.8.4 Texture Area Classification

An image is a collection of pixels with different intensities, where those pixels can be grouped into different texture patches to define the information to be presented in an image. The boundaries of different patches define the shape, location, and orientation of the objects that appear in the image. Those boundaries are regarded as edges, where pixels are observed to have similar intensity within the boundary but an abrupt change in intensity across the boundary [32]. For noise-free images, image edges can be easily identified by the application of edge detector. Different gradient operators can be applied to perform edge detection, where the Sobel operator is a simple but effective detector. The Sobel operator is a first-order derivative operator which is the result of a convolution operation on two 1D *gradient filters*, h_n and h_m, where h_n is for the horizontal direction and the h_m is for the vertical direction. With the normalization factor being 2, the 3×3 Sobel operator is given by

$$h_n = \frac{1}{4}\begin{bmatrix} 1 & 0 & -1 \\ 2 & 0 & -2 \\ 1 & 0 & -1 \end{bmatrix} = \frac{1}{4}\begin{bmatrix} 1 \\ 2 \\ 1 \end{bmatrix}\begin{bmatrix} 1 & 0 & -1 \end{bmatrix} = h_m^T. \tag{1.28}$$

The Sobel operator estimates the horizontal and vertical gradients of the center pixel within a 3×3 kernel. The Sobel edge detector can be implemented with the MATLAB function `sobel(f,threshold)` (see Listing 1.8.9), which returns a binary edge image `edge` of the same size as that of the input image `f`, which outlines the locations of the edges. The edge image has 0 as the entry for all pixel locations that are not edges and 255 for all pixel locations that are edges for a `uint8` image. The values in the map are determined by the threshold of the result of the gradient operation, defined by the input parameter `threshold`. Hence, the final edge map may vary depending on the user-defined `threshold`.

Listing 1.8.9: Sobel edge image extraction.

```
function edge = sobel(f,threshold)
  hn = [1 2 1; 0 0 0; -1 -2 -1];
  hm = hn';
  gn = conv2(double(f), hn, 'same');
  gm = conv2(double(f), hm, 'same');
  s = sqrt(gn.*gn + gm.*gm);
  edge = uint8((s>threshold)*255);
end;
```

To make the discussion complete, let us also go through the MATLAB code of the function `tmap`, which is intuitive for the purpose of loading the workspace parameters and calling the MATLAB built-in `sobel` function to generate the edge map `s`, as shown in Listing 1.8.10.

Listing 1.8.10: Texture map (tmap).

```
function s=tmap(varargin)
  v = varargin{1};
  [sizev] = size(v); % input image size
  if (nargin) == 1
    th = 100; % default threshold value
  else th = varargin{2};
  end
    v = double(varargin{1});
    s = sobel(v,th); %finding edges
end
```

The threshold value applied to the output of the Sobel operator can be imported into `tmap` as the second parameter, where the first parameter is the array that contains the image. If the second parameter is missing, it implies the threshold value is not provided. In that case, a default value of 100 will be assigned as the threshold, where the dynamic range of the input image is assumed to be [0, 255].

The function tmap can be applied to assign different weights to the edge and non-edge pixels in the error image when computing the PSNR to simulate the relative importance of different pixels perceived by the HVS. It should be noted that applying different edge detection algorithm in tmap will lead to minor differences in the result. The Sobel edge detector adopted by tmap is also adopted by the International Telecommunication Union (ITU) [45] for the edge peak signal-to-noise ratio (EPSNR). Without loss of generality, assume the weight w is assigned to the edge pixels, and $1 - w$ to the non-edge pixels. The edge error image is given by

$$e_{edge}[m,n] = (e[m,n])^2(s[m,n](2w-1) + (1-w)), \tag{1.29}$$

where $s[m,n]$ is the edge map of the noise-free reference image. Besides considering the HVS sensitivity difference toward edge and non-edge pixels, the EPSNR further considered the actual contrast of the denoised image by making use of the peak intensity pixel value in the computation of the objective metric instead of the highest possible pixel value as in Equation 1.22. The MATLAB source code Listing 1.8.11 computes the EPSNR.

Listing 1.8.11: Edge peak signal-to-noise ratio (EPSNR).

```
function epsnr_value=epsnr(g,v,w,th)
  e = imageerr(g,v);
  s = tmap(v,th);
  [M,N]=size(g);
  eedge = (e.^2).*((s)*(2*w-1)+(1-w));
  msee = sum(eedge(:))/(M*N);
  epsnr_value = 10*log10(double(255^2/msee));
end
```

where the threshold th is the fourth input parameter in Listing 1.8.11. The matrix eedge is the error image e_{edge}. As you may have noticed from the MATLAB function epsnr, the mean squares edge error is normalized not by the image size, but by the number of pixels that are declared as edge pixels in $s[m,n]$. Similar to PSNR, the higher the EPSNR, the less distortion will be observed on the image edges, and thus, the better the perceived image quality. Finally, it must be pointed out that the EPSNR result is deeply affected by the threshold th, which is the threshold value applied to the gradient results obtained by the Sobel filter to decide whether each pixel location is edge or non-edge pixel. The threshold value should be determined by the local contrast of the image; therefore, a global threshold might not produce good edge detection results and could cause EPSNR to be biased. The following will discuss the structure similarity metric which applies localized analysis to evaluate the difference between the two images.

1.9 Structural Similarity

Noise would corrupt the image in various ways: by altering the original features residing in the homogeneous or the non-homogeneous regions, depending on the nature of the noise, which will result in a different visual appearance between the original image and the noisy image. Figure 1.15 shows selected parts of the *Sculpture* image in the homogeneous region, the structural texture-rich region, and the detail feature-rich region to illustrate the image corruption by different types of noises, where AWGN with $\sigma_n = 50$ and SAP are chosen for comparison. Let us consider the homogeneous area at the *back* of the sculpture, as defined in Figure 1.14. It can be observed that both the AWGN and SAP brought some grain structures to the smooth surface. While natural image can be modeled as a 2D locally stationary Gaussian signal [34], which is in the same nature as that of AWGN, thus AWGN does not alter the image structure much in that particular region but just slightly shifting the brightness. However, the SAP is an impulse noise, which alters the image structure, resulting in annoying artifacts in the region. The *mesh* shows the structural texture of highly repetitive patterns, which has a collection of dominant high-frequency components due to the abrupt pixel contrast of the patterns that appeared in the image spectrum. The dominant components mask both the AWGN and SAP, such that we do not observe significant impact from these noises.

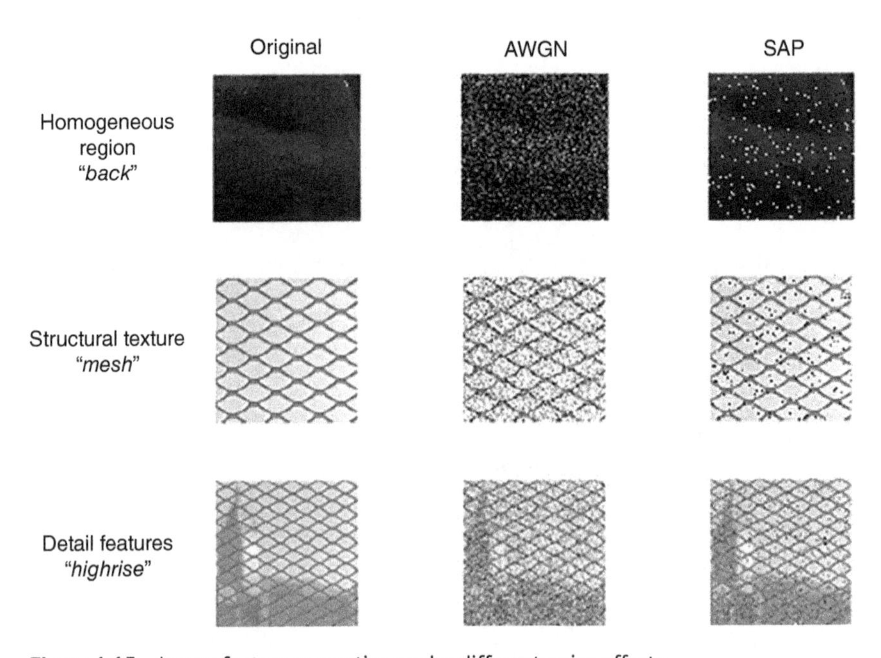

Figure 1.15 Image feature corruption under different noise effects.

Therefore, the lines of the meshes can still be clearly seen in both the AWGN and SAP corrupted cases. However, when we consider the detail feature-rich region, like the *highrise* as shown in the same figure, where this region comprises a blend of patterns which consists of a wide spectrum of frequency components. The AWGN spreads through the whole spectrum, such that the final distribution of different frequency components will be altered, thus washing out the details outlining the *highrise* in the background. While the SAP is impulsive and affects part of the spatial region, thus it has less significant impact on the corrupted image.

We can see that an image would be corrupted in different extent depending on its contents and the nature of noise. The choice of the denoising processes are image- and noise-dependent. Moreover, the way human vision perceives and compares the corruption and the restoration after the denoising process are object-dependent. It is not fair to justify the performance of a denoising method merely by objective measurement, where all pixels in the images are considered and compared in the same fashion without considering the effect of the HVS. The EPSNR is a good start to render the effect of image edges, which are the major attribute that human vision is sensitive to and making use of them to identify objects in an image into the quality measurement. The EPSNR adopts Sobel filter-based edge detection to capture the edges in the image; nonetheless, there are few drawbacks on the application of EPSNR. First, it is a pointwise measurement which does not consider the structure of the image feature. Human vision is more sensitive to image feature that outlines different objects in an image. Second, the accuracy of the edge map is greatly depending on the kernel size of the detection in pointwise basis. Lastly, the edge map detection is a pure luminance comparison, where a human vision-sensitive factor, the contrast, of an image feature is being ignored. Knowing that the luminance and contrast of the image observed by HVS is not a pointwise process but through a small localized region. Therefore, it will be critical to convert the pointwise operation to a localized small image region in the objective quality metric.

To render the perception of luminance, contrast, and structure by human vision in the quality measurement, a variety of HVS-compatible objective quality metrics are proposed for denoised image quality evaluation [20, 51]. Among those reported metrics, the SSIM index proposed by Wang *et al.* in [50], is a benchmark metric in literature, which correlates well with the perceptual image quality. The SSIM is obtained as the product of the luminance, contrast, and structural similarity factors between the denoised image (g) and the reference image (v). These factors are obtained with the use of basic statistical parameters like mean, variance, and covariance as

$$SSIM(g, v) = \frac{(2\mu_g\mu_v + C_1)(2\sigma_{gv} + C_2)}{(\mu_g^2 + \mu_v^2 + C_1)(\sigma_g^2 + \sigma_v^2 + C_2)}, \tag{1.30}$$

where C_1 and C_2 are added to provide stability to each factors, such as to prevent the denominator becoming zero and at the same time bounding the metric to be within a predetermined range (in the case of Equation 1.30, the fraction will be in the range of $[-1, 1]$ but not equal to 0), and μ_x and σ_x are the mean and variance of the random variable x, respectively. Note that the statistical features are computed locally in Equation 1.30. However, the images are generally non-stationary with space variant image structures. Therefore, the localized regions applied to compute Equation 1.30 are extracted by sliding window $\mathbb{W}$ to adapt to the space variant image structure. Starting from the top-left corner of the image, a sliding window of size $\mathbb{W}_S \times \mathbb{W}_S$ moves pixel-by-pixel horizontally and vertically through all the rows and columns of the image until the bottom-right corner is reached. At the kth step, the local quality index $Q_k = SSIM$ is computed within the sliding window. As a result, each processed window will assign an SSIM value at the corresponding pixel coordinate located at the center of the processing window. This forms an SSIM map of the SSIM value for each pixel of the denoised image under concern. If there are a total of K steps, then the overall quality index Q is the *mean structural similarity* (MSSIM) given by averaging all the results obtained in the K steps.

$$MSSIM(g, v) = \frac{1}{K}\sum_{k=1}^{K} SSIM_k(g, v). \tag{1.31}$$

It is vivid that the dynamic range of both SSIM and MSSIM are $[-1, 1]$. The best value 1 can be achieved if and only if $v = g$ for every pixel. The lowest value -1 occurs when $v = 2\mu_g - g$ for every pixels with μ_g being the mean pixel intensity of the image g.

Listing 1.9.1: MSSIM.

```
function [Q,map]=mssim(g,v)
  w = fspecial('gaussian', 11, 1.5);
  w = w/sum(sum(w)); % normalize the filter DC gain
  K = [0.01 0.03]; % default settings
  L = 255; % 8 bit grayscale image
  C1 = (K(1)*L)^2; C2 = (K(2)*L)^2;
  g = double(g); v = double(v);
  mg = filter2(w,g,'valid'); % localized mean by Gaussian filtering
  mv = filter2(w,v,'valid');
  mgs = mg.^2; %g^2
  mvs = mv.^2; %r^2
  mgv = mg.*mv; %gv
  sgs = filter2(w,g.^2,'valid')-mgs; %sigma_g^2 = w(g^2)-g^2
  svs = filter2(w,v.^2,'valid')-mvs; %sigma_r^2 = w(r^2)-g^2
  sgv = filter2(w,g.*v,'valid')-mgv; %sigma_gv = w(gv)-gv
  num1 = 2*mgv + C1; % 2gr + C1
```

```
    den1 = mgs + mvs + C1; % g^2 + r^2 +C1
    num2 = 2*sgv + C2; % 2*sigma_gv +C2
    den2 = sgs + svs + C2; % sigma_g^2 + sigma_r^2 + C2
    map = (num1.*num2)./(den1.*den2);
    Q = mean2(map);
end
```

The MATLAB Listing 1.9.1 implements Equation 1.30 with the sliding window of a 11×11 Gaussian window with unit gain and $\sigma = 1.5$, $C_1 = (K_1 L)^2$, and $C_2 = (K_2 L)^2$, where $L = 255$ is the dynamic range of the pixel intensity for an 8-bit grayscale image. The Gaussian window is chosen instead of other window functions because it can avoid blocking effect which is predominant in windowed local spatial analysis. The readers are referred to [32] and [50] for a throughout study of the SSIM metric, while in this book, we shall take it for granted and apply SSIM in a similar fashion as all other quality metrics presented in Section 1.8.

1.10 Brightness Normalization

The pixel intensity of a `uint8` image is bounded in [0,255]. To maintain the proper representation and display of the digital image, we need special operation for any overflow or underflow of pixel intensities. In Section 1.5.2, a MATLAB function `imgtrim` (see Listing 1.5.4) has been introduced to handle these circumstances by truncating all negative pixel values and capping all pixel values greater than 255 to 255. This function serves as a simple post-correction any numerical errors caused by noise injection or the later denoising process. Nonetheless, this correction is nonlinear, which may not be suitable for quality evaluation. The MATLAB function `brightnorm`, as shown in Listing 1.10.1, normalizes the pixel intensities in the restored image `g` to give the brightness normalized image `ng`, which has the same dynamic range as that of the original image `f`.

Listing 1.10.1: Brightness normalization.

```
function ng = brightnorm(f,g)
  [Lmax MaxInd] = max(f(:));
  [mmax, nmax] = ind2sub(size(f),MaxInd);
  [Lmin MinInd] = min(f(:));
  [mmin, nmin] = ind2sub(size(f),MinInd);
  fmax = Lmax;
  fmin = Lmin;
  gmax = max(max(g(1:2:end,1:2:end)));
  gmin = min(min(g(1:2:end,1:2:end)));
  if gmin<0
    gmin =0;
```

```
    end;
    g(g<0)=0;
    scale = (fmax-fmin)/(gmax-gmin);
    ng = (g-gmin)*scale+fmin;
    ng = ng*(mean(mean(f))/mean(mean(ng)));
end
```

Please note that `brightnorm` altered the pixel values nonlinearly. As a result, image processed by `brightnorm` will not be suitable to be evaluated by any of the analytical method listed in this chapter, except the SSIM.

1.11 Summary

Noises are the random elements incorporated into the image capturing and processing results, which cannot be easily controlled, and may be dependent on, or independent of, the image content. Noise is the key problem in 99% of cases where image processing techniques either fail or further image processing is required to achieve the required results. Therefore, a large part of image processing algorithms is dedicated to increase the robustness of the algorithm through the application of some kind of denoising process.

In this chapter, we have introduced the analytical model of image noises, briefly presented how to estimate the noise power in an image through local statistics, and finally briefly discussed the idea of image quality measurement, which will help to evaluate the performance of a denoising algorithm. In our discussions, we have particularly chosen full-reference quality measurement, where the original (noise-free) image is considered to be available a prior for comparison. Both objective and subjective image quality measurements have been discussed. The objective quality measurement, in contrast to the subjective measurement, is conducted by the image quality metric which counts the difference between the original image and the distorted image. The MSE is the most common objective quality measurement metric that is widely used in the literature, and it forms the basis of other objective quality metrics, such as the PSNR and SSIM. The numerical nature of objective quality measure helps to provide a fairly easy performance ranking system, which plays an important role in comparing the performance of a variety of image denoising applications. However, let us not forget that in most applications, the human observer will be the final judge of the image denoising performance. Therefore, subjective quality evaluation is equally important in measuring the image denoising algorithm performance. In this chapter, we have discussed the image features that should be focused on when subjective image quality evaluation is applied. It will be difficult to mathematically quantify subjective measure. On the other hand, descriptive performance evaluations are usually applied, and in particular, over selected image features. Later chapters will make use of these evaluation methods to discuss the performance of various image denoising algorithms presented in the book, and quantify the possible performance improvement methods for the image denoising algorithm in concern.

Exercises

1.1 MATLAB experiment:

1. View the data type of the grayscale image "sculpture.tif".
2. Convert the data type of grayscale image to `double`.

1.2 The ITU-R recommendation BT.601 [44] defines $k_b = 0.114$ and $k_r = 0.299$, and further constrain Y in the range of $[0, 255]$ with the RGB in the same range. Derive the equations that convert RGB to YCbCr, and from YCbCr to RGB, and implement the conversion function in MATLAB.

1.3 There are two main groups of methods to convert color images to grayscale images: luminosity methods and color-altering methods. Luminosity method is based on the brightness of the color while color-altering method is according to the color contrast of the image. Develop a MATLAB program that performs color-to-grayscale image conversion for the *Sculpture* image to grayscale image by

1. Luminosity method with the following conversion formulas:
   (a) Green weighted

$$Y = 0.21R + 0.62G + 0.17B. \tag{1.32}$$

   (b) Red weighted

$$Y = 0.62R + 0.21G + 0.17B. \tag{1.33}$$

   (c) Blue weighted

$$Y = 0.62R + 0.17G + 0.21B. \tag{1.34}$$

2. Average method

$$Y = \frac{R + G + B}{3}. \tag{1.35}$$

3. Lightness method

$$Y = \frac{1}{2}(\max(R, G, B) + \min(R, G, B)). \tag{1.36}$$

Compare and comment on the obtained grayscale images from the above five formulas.

1.4 Develop a MATLAB program to load the color *Sculpture* image and convert the color image into grayscale image using
1. MATLAB image processing library function `rgb2gray`,
2. $y = \frac{r+g+b}{3}$.

Are the two images the same? Explain your observation.

1.5 Apply the salt and pepper noise function `sapnoise` to your selected natural image, and study the histogram with respect to different `nd`. Comment on your observation with respect to `nd`.

2

Filtering

One of the most common problems encountered in the area of image processing is noise removal. Any image captured by conventional physical devices (either cameras or digital cameras, etc.) will pick up noise from a variety of sources. The noise usually has a negative aesthetic effect on the human eyes. Moreover, it is best to remove the noise to maximize the performance of subsequent uses of the digitized image such as in computer vision, classification, recognition, etc. Hence, a large number of techniques have been presented in literature to address this problem, ranging from the typical lowpass filter that convolutes the original image with a predefined mask, to the more advance and also more complex methods involving patch matching, differential equations, and approximation theory.

In practice, an image may be degraded by various types and forms of noise. The most common type of noise is the additive one. As shown in Figure 1.12, the degradation process is modeled by an additive noise term, η, which is added to the input image υ, also known as the noise-free image and the reference image, to produce a degraded image f. Given this noisy observation, along with some knowledge of the additive noise, certain restoration technique can be applied to estimate the noise-free image where the estimated image is known as the denoised image g. Getting back the noise-free image υ by observing f alone is impossible. Indeed this is an ill-posed problem. Fortunately, it has been demonstrated in Figure 1.15 that the HVS has certain noise masking capability depending on the image structure. Therefore, even if the denoised image g is not the same υ. As long as g is close to the original image, the human eye will consider both images to be the same.

Choosing which denoising technique to be used poses a number of problems. Consider the noisy step image in Figure 2.1(b), which is an AWGN corrupted image of Figure 2.1(a) with $\sigma_\eta = 50$. It is easy to associate its noise-free image is Figure 2.1(c) instead of Figure 2.1(a). To determine the noise-free image associated with Figure 2.1(b) requires determining the pixel intensity on both sides of the step will suffice. Because the eye is capable to determine the average pixel intensity over a small localized region, which is also known as homogenous masking. Therefore,

Digital Image Denoising in MATLAB, First Edition. Chi-Wah Kok and Wing-Shan Tam.

Companion website: www.wiley.com/go/kokDeNoise

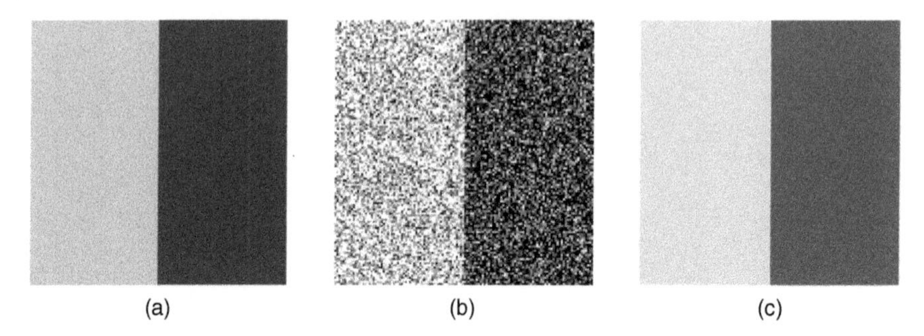

Figure 2.1 (a) Noise-free step edge image with pixel intensities at 128 and 58; (b) AWGN corrupted step edge image in (a) with noise variance $\sigma_\eta = 50$; and (c) noise-free step edge image with pixel intensities at 193 and 93.

the noise fluctuations over the small localized region will be averaged. When the noise has zero mean, the average should remove the noise and recover the intensity of the noise-free image. However, since the region observed by human eye is small, it is most likely that the mean value of the noise within this small region is biased. As a result, the image will be observed to appear with a different tone when compared to that of the noise-free image. This is the reason why the noise-free image of Figure 2.1(b) looks like Figure 2.1(c), instead of Figure 2.1(a). On the other hand, it should also be observed that the averaging effect of the human eye is an effective device to remove the image noise. Consider a spatial frequency point of view explanation on the averaging effect, which is the recovery of pixel intensities from white noise by lowpass filtering. The eye is a device that performs the lowpass filtering. The denoising efficiency of lowpass filtering comes from the particular energy distribution of natural image in the frequency domain. The noise-free signal spectrum of natural image is usually concentrated in low frequencies. The radial spectrum of a natural image is commonly observed to be an exponential decay function $1/r^\beta$ with radial frequency r and decay parameter β, where β is observed to be close to 2 for most of the natural images. Figure 2.2 shows the spectral magnitude of the image *Sculpture* corrupted by AWGN with $\sigma_\eta = 50$ in radial frequency domain. It can be observed that the natural image *Sculpture* decays exponentially with r, while the AWGN has a flat spectrum, close to a constant with respect to r. The image spectrum has higher power than that of the noise spectrum when $r < r_0$. Therefore, a mere lowpass filtering of the noisy image can already improve the signal-to-noise (SNR), when the passband of the lowpass filter is close to r_0. The search for the best denoise filter can be formulated as an optimization problem of

$$\min_h \|h \otimes f - v\|^\alpha, \tag{2.1}$$

Figure 2.2 The average spectral power at radian frequency $r = \sqrt{\omega_x^2 + \omega_y^2}$ of the *Sculpture* image and additive Gaussian noise. At radian frequency r_0 and larger, the noise power will dominate spectrum.

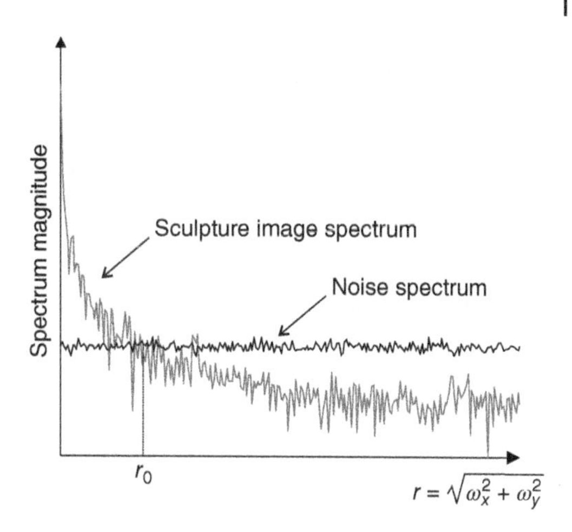

where α is the order of the norm function. The popular choice of α is 2, such that Equation 2.1 minimizes the power of the differences between the filter denoised image and the noise-free image. Other choice of α can be applied, and other distance functions can also be applied to replace the norm operator. The study of such denoise filter is the main topic of the present chapter.

2.1 Mean Filter

The additive noise spreads almost constantly throughout the entire image spectrum but only those lying in the high-frequency spectrum would be in concern, as the noise energy dominates the high-frequency spectrum and is not masked by the original image signal (see Figure 2.2). It is reasonable to consider the additive noise in the high-frequency region can be blocked or filtered using averaging technique to recover the best-estimated image g. The simplest filter that can be used is an ideal lowpass filter, which has a constant passband starting from the DC component (in the center of the *Fourier Transform* (FT)) up to a particular cut-off frequency r_0. The transfer function of the ideal filter is given by

$$H(u, v) = \begin{cases} 1, & \text{if } R(u, v) \leq r_0, \\ 0, & \text{if } R(u, v) > r_0. \end{cases} \tag{2.2}$$

where $R(u, v)$ is the Euclidean distance between a point of coordinate (u, v) from the origin on a frequency plane (the FT domain in this case), and r_0 is a nonnegative value, referred to as the cut-off frequency and cut-off radius. As a result, a simple definition on $R(u, v)$ can be

$$R(u, v) = \sqrt{u^2 + v^2}. \tag{2.3}$$

Figure 2.3 Ideal lowpass filter: (a) the 512 × 512 filter in 2D view in frequency domain with $r_0 = 50$, (b) the denoised image *Sculpture* using ideal lowpass filter (noisy image: Figure 1.8(b), PSNR = 20.096 dB), and (c) the zoomed-in portions of the denoised image.

Figure 2.3(a) shows the 2D plot of an ideal lowpass filter with $r_0 = 50$, and the spectral width being 512, where the white region in the center is the passband and the dark region that encloses the stopband of the filter. The MATLAB code to implement the ideal filter and its frequency response plot in Figure 2.3(a) is shown in Listing 2.1.1.

Listing 2.1.1: Ideal filter.

```
% Part I: define the frequency plane
[M,N] = size(f);
u = 0:(M-1); v = 0:(N-1);
idx = find(u>M/2);
u(idx) = u(idx)-M;
idy = find(v>N/2);
v(idy) = v(idy)-N;
[V,U] = meshgrid(v,u);
R=sqrt(U.^2+V.^2);

% Part II: define the filter
r0 = 50;              % define the cutoff point
Ht = double(R<=r0); % filter defined by the distance to cutoff
H1 = Ht(1:M/2,1:N/2);
H3 = Ht(M/2+1:end,1:N/2);
H2 = Ht(1:M/2,N/2+1:end);
H4 = Ht(M/2+1:end,N/2+1:end);
H = [H4 H3; H2 H1];
figure;imshow(255*uint8(H));
```

Part I of the code defines the frequency plane for the ideal filter `H`. The size of the filter has to be consistent to that of the noisy image `f`. The last six lines of the code in Part II renders the proper display of the filter with the passband centered at the origin.

The convolution of the filter to the noisy image ($h \otimes f$) in spatial domain is the multiplication of their spectra in frequency domain, given by

$$G(u, v) = H(u, v)F(u, v), \tag{2.4}$$

where $F(u, v)$ is the Fourier transform of the image being filtered and $H(u, v)$ is the filter transform function. The MATLAB implementation is shown in Listing 2.1.2. The function `fft2(f)` returns `ft`, which is the 2D discrete Fourier transform (DFT) of the noisy image f. The denoised image spectrum `gt` can be obtained by simple element-by-element multiplication operation between the filter spectrum `H` and the noisy image spectrum `ft`. The function `ifft2(gt)` returns `g`, which is the 2D discrete inverse Fourier transform of the denoised image spectrum G. It should be noted that the output of the `ifft2` function is conjugate symmetric, which has one half of its spectrum in positive frequencies and the other half in negative frequencies. Only the positive half is considered in the operation, such that `real` command is applied to rectified the final output.

Listing 2.1.2: Ideal filter.

```
[M,N] = size(f);
ft = fft2(double(f));
gt = H.*ft;
g = uint8(real(ifft2(double(gt))));
```

Figure 2.3(b) shows the recovered images of the AWGN corrupted *Sculpture* image (Figure 1.8(b)) using the ideal filter, where its PSNR is 20.096 dB. The ideal filter presents a sharp transition from the passband to the stopband. As a result, the multiplication of the filter to the noisy image in the frequency domain is in fact a hard thresholding operation that all frequency components locating outside the cut-off frequency will be dropped. This method is simple and it provides a tight control of the cut-off frequency of the filter. Nevertheless, it comes with the drawback that the output image suffered from the Gibbs phenomenon [48], where ripple-alike pixel intensity variations will be observed across edges of all objects in the denoised image. Figure 2.3(c) shows the zoomed-in portions of the recovered image (b). In the *bench* region, it can be observed that the low-frequency components in this homogeneous region are well preserved, which demonstrates the efficiency of the ideal filter. However, it can be observed in those image regions that contain high-frequency components, such as the sharp edge across the *ear* and the texture region formed by the metal meshes behind the *book & glass*, were both blurred, washed out, and ringing noises are observed in the regions surrounding these edges. This artifact is caused by the ripple-alike spatial domain response of the ideal filter, where the spatial domain response of the ideal filter is an infinite sinc function, and the ripple alike noises are also known as the Gibbs phenomenon.

When there is no prior information on the noise nor the noise-free statistic, *moving average filter* is the popular alternative, which provides a finite length filtering in the spatial domain that helps to suppress high-frequency noise and alleviate ringing problem caused by Gibbs phenomenon. The output of the *moving average* filter removes the instantaneous noise and replace the intensity of the noisy pixel with the average of the intensity of the pixels within the filter window. *Mean filter*, also known as *box filter*, is one of the simplest implementation of moving average filter. *Mean filter* takes the average of the pixel intensities of all neighboring pixels covered by the filter window $\mathbb{W}$ with size of $\mathbb{W}_M \times \mathbb{W}_N$, where the mean filter is given by

$$h_{ave}[m,n] = \begin{cases} \frac{1}{\mathbb{W}_M \times \mathbb{W}_N}, & 0 \le m < \mathbb{W}_M \quad \text{and} \quad 0 \le n < \mathbb{W}_N \\ 0, & \text{otherwise} \end{cases} \tag{2.5}$$

This is equivalent in applying the same weightings to all neighboring pixels regardless of their distance to the center pixel. The intensity of the center pixel value would be replaced by the output of the filter. The window is also known as *kernel*, which is a pixel block moving along row then column in the image. The size of the kernel can be adjusted to achieve the averaging effect in different extent. Listing 2.1.3 shows how to make use of the MATLAB built-in function to implement the mean filter.

Listing 2.1.3: MATLAB built-in function for mean filtering.

```
>> w=3;
>> have=fspecial('average',w);
>> g=filter2(have,f,'same') ;
```

The function `fspecial` creates a 2D filter `have` of the specified filter type, where the first parameter `'average'` specifies the filter to be created is a mean filter and the second parameter w specifies the kernel size, where the kernel size in this example is 3×3. It should be noted that the default kernel size is 3 if the second parameter is omitted in the `fspecial` command for the filter type `'average'`. The function `filter2` performs the 2D convolution between the input signal `f` and the filter `have` rotated by 90°, where the output of this function is in the same size as that of the central part of the input signal `f`. The `have` obtained from Listing 2.1.3 is equivalent to

$$h_{ave}(m,n) = \frac{1}{3^2}\begin{bmatrix} 1 & 1 & 1 \\ 1 & 1 & 1 \\ 1 & 1 & 1 \end{bmatrix}, \tag{2.6}$$

which can be obtained from Equation 2.5 with both $\mathbb{W}_M$ and $\mathbb{W}_N$ equal 3. Therefore, we can always define the filter kernel by simple matrix array manipulation using Equation 2.5 instead of using the built-in function `fspecial`. Listing 2.1.4 implements the h_{ave} filter in Equation 2.6 is listed in the following.

Listing 2.1.4: Image denoising by a 3×3 mean filter.

```
>> have=[1 1 1; 1 1 1; 1 1 1]./(3*3);
>> g=filter2(have,f);
```

Figure 2.4(a) shows the frequency response of the mean filter in Equation 2.6. It can be observed that a rectangular mean filter in spatial domain (in one of the spatial axis) is a sinc function in frequency domain. The passband in the low-frequency region is narrow. However, the high-frequency region is occupied by ripples, therefore, parts of the high-frequency components can pass through the filter due to the non-zero filter components in the high-frequency region, which is in great contrast to the filtering performance of the ideal filter.

The MATLAB functions in Listings 2.1.3 and 2.1.4 will yield the same filtered image g. Figure 2.4(b) shows the restored image from the noisy image with AWGN of $\sigma_\eta = 50$ (the noisy image is shown in Figure 1.8(b)) using a 3×3 mean filter, in which the PSNR of the restored image is 20.41 dB. The AWGN has zero mean. As a result, it is expected that the mean filter would average the noise within the kernel to zero, thus improving the PSNR when compared to that of the ideal filter case. The noise variance of the mean filtered image can be obtained as

$$\sigma_{g,\eta}^2 = \frac{1}{\sqrt{\mathbb{W}_M \times \mathbb{W}_N}} \sigma_\eta^2, \tag{2.7}$$

where $\mathbb{W}_M \times \mathbb{W}_N$ is the size of the mean filter kernel. It can be observed that the power of the noise can be reduced by increasing the kernel size of the mean filter. In the subjective aspect, let us consider the zoomed-in portions of the recovered image Figure 2.4(b) and compare them with those shown in Figure 2.3(c). We can see that the mean filter can better preserve the high-frequency components in the image (the meshes are less washed out around the *ear* and *book & glass*) when compared to that of the ideal filter. However, the noisy components

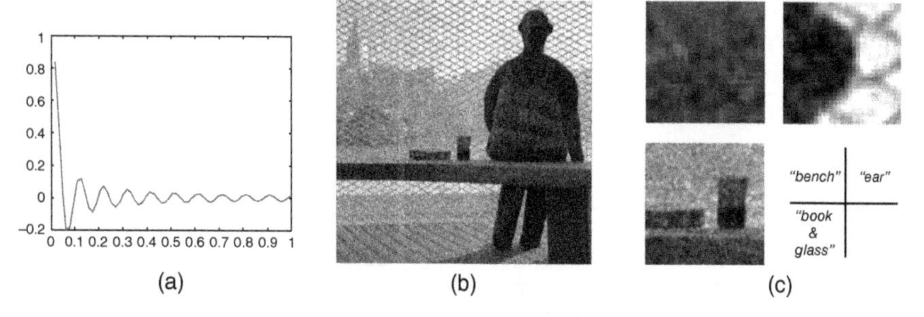

Figure 2.4 Mean filter: (a) plot of a mean filter in frequency domain, (b) the denoised image *Sculpture* using a 3×3 mean filter kernel applied to AWGN corrupted image (noisy image: Figure 1.8(b), PSNR = 20.41 dB), and (c) the zoomed-in portions of the denoised image.

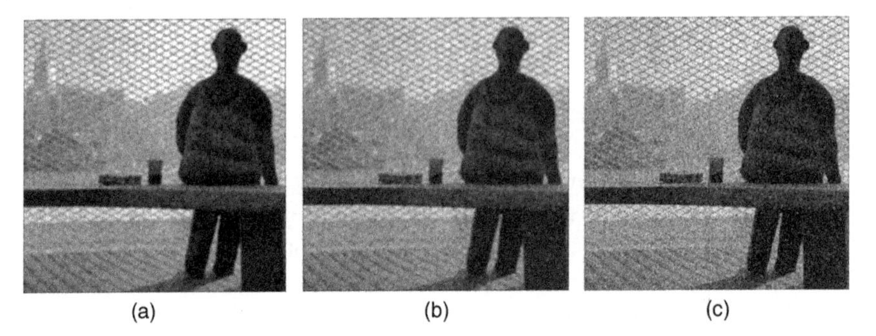

(a) (b) (c)

Figure 2.5 Effect of mean filter kernel size versus different sources of noise: (a) a 5×5 mean filter kernel applied to AWGN corrupted *Sculpture* image with $\sigma_\eta = 50$ (noisy image: Figure 1.8(b), PSNR = 20.26 dB), (b) a 5×5 mean filter kernel applied to AWGN and SAP mixed corrupted image (noisy image Figure 1.11(a), PSNR = 19.2438 dB), and (c) a 3×3 mean filter kernel applied to AWGN and SAP mixed corrupted image (noisy image Figure 1.11(a), PSNR = 19.2422 dB).

in the homogeneous region (at the *back*) are more apparent, which is due to the non-zero stopband in the high-frequency regions, where some of the noises in the high-frequency region are not able to be filtered out.

To further investigate the noise reduction selectivity of the mean filter, let us consider the images in Figure 2.5 which are the mean filtering result of the *Sculpture* image corrupted with different kind of noises, and denoised with mean filter with different kernel size. It can be observed from Figure 2.5(a) that the denoised image of the AWGN corrupted *Sculpture* image with $\sigma_\eta = 50$ in Figure 1.8(b) with a 5×5 mean filter is inferior to that obtained by the 3×3 mean filter (see Figure 2.4(b)). At the same time we can also observe that the larger the filter kernel size, the more blurred the denoised image. This is because the larger the filter kernel size, the more disperse the spatial domain response of the filter kernel which makes it further apart from that of an impulse function and finally the blurring effect. However, if we accept blurring against noise reduction, the mean filter with large kernel size is able to do a better job for image corrupted with mixed AWGN and SAP. Showing in Figure 2.5(b) and (c) are the denoising results for the *Sculpture* image corrupted by SAP with density 0.5, and AWGN with $\sigma_\eta = 50$ obtained by mean filters with kernel size 5×5 and 3×3, respectively. The 5×5 mean filter effectively removed the SAP, while the SAP passes through 3×3 mean filter. On the other hand, the denoised image obtained from 5×5 mean filter is more blurred than that obtained by the 3×3 mean filter. Nevertheless, being able to alleviate the SAP problem helps in obtaining a high PSNR than the reduction of AWGN, where we can observe that from the PSNR of the denoised images, which are 19.2438 and 19.2422 dB obtained via mean filter with kernel size 5×5 and 3×3, respectively.

2.1.1 Gaussian Smoothing

The Gaussian smoothing is another spatial domain denoising technique. It is as effective as the mean filter to remove AWGN. The Gaussian smoothing filter, also known as Gaussian filter, is a lowpass filter with cut-off frequency at r_0. Generally, r_0 will be chosen at a particular frequency where the noise power dominates the spectrum beyond that particular frequency. As a result, if a lowpass filter at r_0 is applied to the Gaussian noise-corrupted image, all noise dominated high-frequency components of the corrupted image will be removed just as what the ideal filter does. However, the lowpass filter will also remove the high-frequency components of the original image, and hence the recovered image will be blurred. On the other hand, unlike the ideal filter, the Gaussian filter defined by

$$h_G(m,n) = \frac{1}{2\pi\sigma^2}e^{\frac{-(m^2+n^2)}{2\sigma^2}} \xleftrightarrow{DFT} H_G(u,v) = \frac{1}{2\pi\sigma^2}e^{\frac{u^2+v^2}{2\sigma^2}}, \tag{2.8}$$

where σ is the passband roll off factor of the Gaussian filter has a smooth transition band as can be observed from the 2D frequency response plotting in Figure 2.6(b). When h_G is considered as a distribution function, σ will be the standard deviation. It should be noted that the smooth passband roll off of the Gaussian filter helps to alleviate the Gibbs phenomena. Furthermore, the Gaussian filter has full symmetry in the spatial frequency domain, as observed in Figure 2.6(c), which shows the same characteristics as that of the ideal filter in Figure 2.3. This spatial frequency property helps to ensure that the cut-off frequency is universal along any spatial direction, and hence provides better denoising results. Consider the Gaussian filter with kernel size of 5×5 is applied to denoise the AWGN corrupted *Sculpture* image in Figure 1.8(b). The filtered image is shown in Figure 2.7(a). It can be observed that the Gaussian-filtered image is blurred and have almost all AWGN removed. At the same time, no ringing noise is observed in contrast to that obtained by the ideal filter. Furthermore, the frequency response of the Gaussian

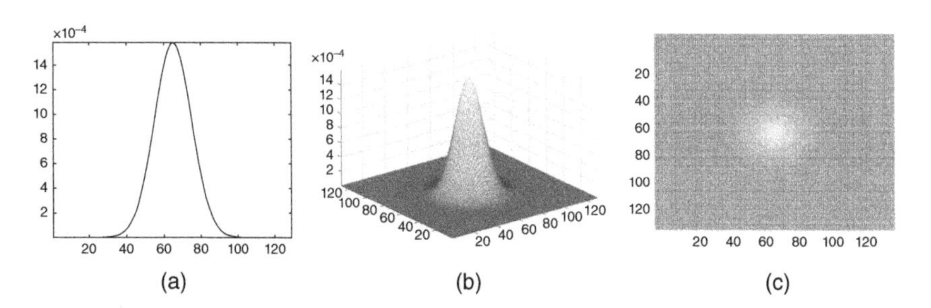

Figure 2.6 (a) Spatial frequency response of the Gaussian filter along m-axis with kernel size 5×5; (b) 3D view of the frequency response of the Gaussian filter in (a); (c) aerial view of the frequency response of the Gaussian filter.

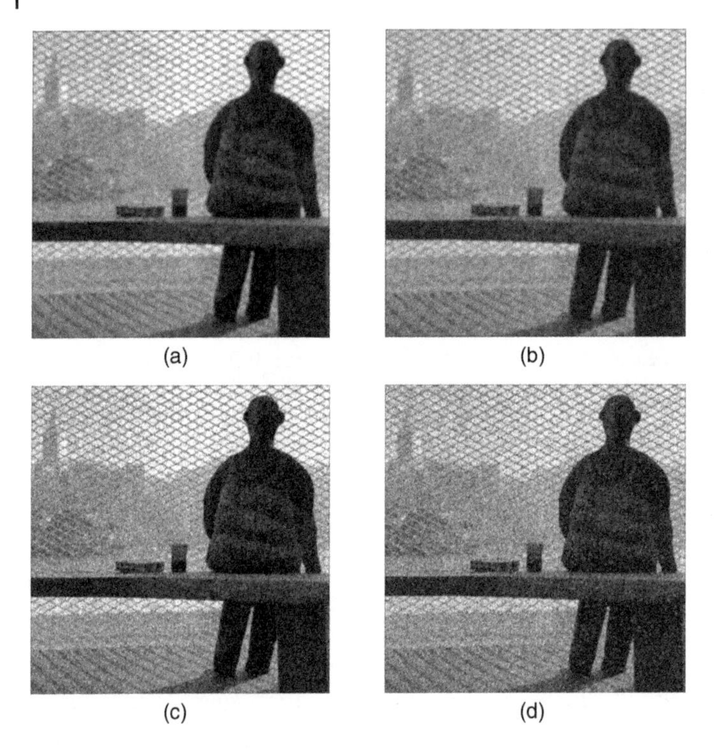

Figure 2.7 Effect of the size of a Gaussian filter with identical passband roll off factor $\sigma = 5$ versus different sources of noise: (a) a 5×5 Gaussian filter applied to an AWGN corrupted image (noisy image: Figure 1.8(b), PSNR = 20.33 dB), (b) a 5×5 Gaussian filter applied to an AWGN and SAP mixed corrupted image (noisy image Figure 1.11(a), PSNR = 19.30 dB); (c) a 3×3 Gaussian filter applied to an AWGN corrupted image (noisy image: Figure 1.8(b), PSNR = 20.43 dB), and (d) a 3×3 mean filter kernel applied to an AWGN and SAP mixed corrupted *Sculpture* image (noisy image Figure 1.11(a), PSNR = 19.26 dB).

filter is much sharper than that of the mean filter. As a result, the blurring effect of the denoised image obtained by Gaussian filter is observed to be not as bad as that obtained by the mean filter.

The MATLAB has a built-in Gaussian filter kernel generation function, as shown in Listing 2.1.5. The function `fspecial` creates a 2D filter specified by the filter identifier, and in our example `guassian` is supplied to define the required filter. For Gaussian filter, the second and the third input parameters specify the filter size and the value of the σ in the Gaussian filter. In our example, the filter size is chosen to be 5×5 and $\sigma = 5$. `h` is the spatial response of the Gaussian smoothing filter and is applied to the noisy image `f` through 2D convolution, where in MATLAB we use the function `imfilter` to filter the noise-corrupted image `f` and give the recovered image `g`.

Listing 2.1.5: Gaussian filter.

```
>> f = imnoise(r,'gaussian',0,0.1);
>> h = fspecial('gaussian',5,5);
>> g = imfilter(f,h);
```

Without going into MATLAB built-in function, a $k \times k$ Gaussian kernel can be generated by `gaussiankernel(k,sigma)` in Listing 2.1.6,

Listing 2.1.6: Gaussian filter kernel generation.

```
function h = gaussiankernel(k,sigma)
  ind = -floor(k/2):floor(k/2);
  [p,q] = meshgrid(ind,ind);
  h = exp(-(p.^2+q.^2)/(2*sigma^2));
  h = h/sum(h(:));
end
```

The same 5×5 Gaussian filter that applied in Figure 2.7(a) is applied to a mixed AWGN and SAP corrupted *Sculpture* image (see Figure 1.11(a)) to give a denoised image, as shown in Figure 2.7(b). It can be observed that the Gaussian smoothing effectively removed the AWGN in the homogenous regions, while the SAP will be blurred in singular region of the image. When the area corrupted by SAP is a homogeneous area, uneven and unnatural intensity distribution within the localized area around the noisy SAP pixel is observed in the denoised image. When the SAP is located at or around the singular region of the image, the spreading of the SAP will produce masking effect onto the edges of the image objects that are located around the SAP corrupted pixels. This observation suggests that reducing the kernel size can help the Gaussian filter to better locate itself away from the singular regions. The simulation results presented in Figure 2.7(c) and (d) are evidence of the above observation where the Gaussian filter with kernel size 3×3 and $\sigma = 5$ is applied. It can be observed from Figure 2.7(c) and (d) that the images are less blurred when compared to their counterparts shown in Figure 2.7(a) and (b). At the same time, the PSNR of the denoised images in Figure 2.7(c) and (d) have improved PSNR which are 20.43 and 19.26 dB, respectively.

2.2 Wiener Filter

Is it possible to find the best linear filter for image denoising? To answer this question, we shall formulate the problem of finding the best filter h_{opt} as an optimization problem that minimizes the expected mean squares error between the noise-free image and the denoised image as

$$h_{opt} = \min_{h} E[\|(h \otimes f) - v\|^2], \tag{2.9}$$

where $E[\cdot]$ is the expectation operator. The solution to this question is the Wiener filter. Its frequency response is given by

$$H_w(\omega) = \frac{S_\upsilon(\omega)}{S_\upsilon(\omega) + S_\eta(\omega)}, \tag{2.10}$$

where $S_\upsilon(\omega)$ is the power spectral density of the noise-free image υ and $S_\eta(\omega)$ is the power spectral density of the noise η.

The Wiener filter is optimal with respect to the minimum mean squares error (MMSE), space-invariant linear estimator of a stationary noise-free image degraded by additive stationary noise. However, when applied to real-world image denoising problems, it suffers from two main weakness: first, it does not make any distinctions between the treatment of edges and homogeneous due to its space-invariance; second, its performance highly depends on the estimation of the power spectrum of the noise-free image. It is globally optimal for a Gaussian stationary random signal only. The problem is that natural images do not belong to this particular class of signals. There is a considerable amount of literature that investigates efficient way of estimating the noise-free signal power spectral density in more general cases. The most popular approaches impose some parameterized models (generalized Gaussian or Laplacian, fractal-like, etc.) depending on the class of signals to which the noise-free signal is supposed to belong. Due to this difficulty, the Wiener filter is usually outperformed by spatially adaptive and/or non-linear algorithms.

The biggest problem of Wiener filter will be the requirement on the complete knowledge of the signal spectrum, which is almost always not available. To remedy this problem, the noise spectrum is estimated by local mean and variance around each pixel. Such Wiener filter is known as the pixelwise Wiener filter, such that given a window $\mathbb{W}$ of size $\mathbb{W}_M \times \mathbb{W}_N$, a local mean and variance around pixel $[m, n]$ is estimated via

$$\mu = \frac{1}{\mathbb{W}_M \times \mathbb{W}_N} \sum_{[m,n]\in\mathbb{W}} f[m, n], \tag{2.11}$$

$$\sigma^2 = \frac{1}{\mathbb{W}_M \times \mathbb{W}_N} \sum_{[m,n]\in\mathbb{W}} (f[m, n]) - \mu)^2. \tag{2.12}$$

The pixelwise Wiener filter will yield a pixelwise Wiener filtered output as

$$g[m, n] = \mu + \frac{\sigma^2 - \sigma_\eta^2}{\sigma^2}(f[m, n] - \mu), \tag{2.13}$$

where $\sigma^2 - \sigma_\eta^2$ estimates the power of S_υ in Equation 2.10 and σ^2 estimates the power of $S_\upsilon + S_\eta$ in Equation 2.10, and hence Equation 2.13. MATLAB has a built-in function `wiener2` that implements the above pixelwise Wiener filter.

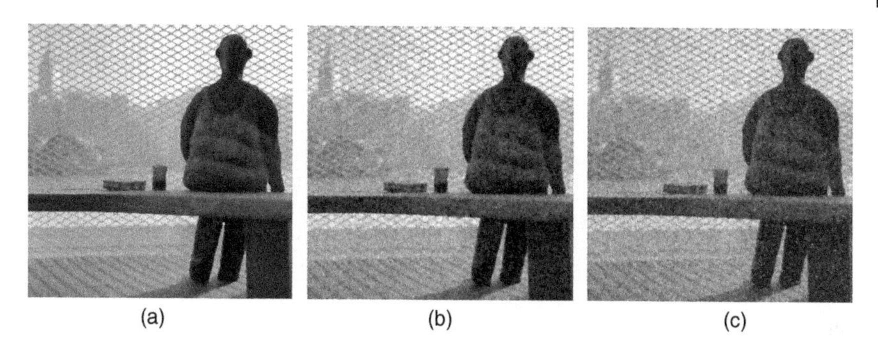

Figure 2.8 Wiener filtering denoised *Sculpture* images of (a) AWGN corrupted image with $\sigma_\eta = 10$ (noisy image: Figure 1.8(a), PSNR = 25.44 dB), (b) AWGN corrupted image with $\sigma_\eta = 50$ (noisy image: Figure 1.8(b), PSNR = 21.39 dB), and (c) AWGN and SAP mixed corrupted image (noisy image: Figure 1.11(a), PSNR = 19.58 dB).

The following MATLAB script makes use of `wiener2` to denoise the noisy image in Figure 1.8(b).

```
>> g = wiener2(f,[5 5]);
```

where the implementation of pixelwise Wiener filter in MATLAB will be an exercise for the reader. The denoised images shown in Figure 2.8 are obtained by pixelwise Wiener filter with local window size of 5×5 on *Sculpture* images corrupted with different noise sources. It can be observed that the Wiener filtering achieves better noise removal objectively, where larger PSNR are observed in all three cases when compared to that of the Gaussian filter cases. This performance gain is due to the sub-optimal solution in the localized window. In other words, the window size would affect the performance of the filtering. We shall leave it as an exercise for the readers. In the subjective performance aspect, the Wiener filtering reduces the blurring artifacts in the denoised images, where we can see the mesh in the background are more clearly preserved. This is because Wiener filter is not a lowpass filter and thus will not eliminate all the high-frequency components in the image, which implies the high-frequency components of the edges, and other textures can pass through Wiener filter and hence the more vivid appearance in the filtered image. In Figure 2.8(a) and (b), we can observe that the Wiener filtering is effective in suppressing the AWGN, when compared to that of the SAP as shown in Figure 2.8(c). It is because the SAP is impulsive rather than a Gaussian process. However, it achieves a better SAP suppression along the image edges, when compared to that of the Gaussian filter case, as the local covariance structure within the filter kernel is considered in the Wiener filtering through the small local window. Though Figure 2.8(a) and (b) are corrupted with AWGN only, the larger noise variance in Figure 2.8(b) results in degraded covariance structure in the kernel, which increases error in the optimization process, and it shows less pleasant visual appearance (with more apparent intensity fluctuations) in the homogeneous regions.

2.3 Transform Thresholding

The lowpass filtering denoise methods that are discussed in Sections 2.1 and 2.2 have demonstrated that the noise components are removed based on their frequency composition. The lowpass filtering removes the noise if the frequency component of the noise is over a particular frequency range, which is the cut-off frequency region as aforementioned. The cut-off frequency can be considered as a threshold in the frequency domain. This thresholding concept can be extended into other image representation domains, e.g. *Discrete Fourier Transform* (DFT) domain, *Discrete Cosine Transform* (DCT) domain, or *Wavelet Transform* (WT) domain, etc., and it is natural to ask the question of whether thresholding in other domains will be able to achieve better denoising result. There will be two questions arisen at this moment: first, which transform should be applied; second, how to determine the associated thresholding parameters. The answer to the first question can be investigated by knowing that the coefficients of an homoscedastic and decorrelated noise remain homoscedastic and decorrelated in any orthogonal transform. As a result, any orthogonal transform should achieve the same performance. However, the choice of orthogonal transform will not only affect the noise, it will also affect the determination of the associated threshold parameters. For example, the DCT coefficients of a Gaussian white noise with variance σ_η^2 remains a Gaussian diagonal vector with variance σ_η^2 in the DCT domain. Thus, a threshold on the coefficients at, say $3\sigma_\eta$ removes most of the coefficients that are only due to noise, assuming the expectation of these coefficients is zero. The sparsity of transform coefficients of natural image represented in the DCT orthogonal basis has provided a very easy and effective coefficient thresholding method to denoise the image. The same property is applied in JPEG to achieve image compression with the decompressed image faithfully resembles the original image. Among all of them, the DCT is well known to be able to provide a near optimal representation of natural signal with high energy compaction, which making the DCT being widely used in many image and video compression algorithms, such as JPEG, MPEG, IUT/T, etc. In this section, we shall focus on denoising in the DCT domain, but the readers should be able to apply the idea studied in this book to other orthogonal transformations.

Image processing in the DCT domain is a block-wise operation, where the transform is applied in each block separately. Listing 2.3.1 shows the MATLAB function `blockdct` for the block-wise operation of DCT on an input image `f` with the block size set at `L` × `L`. The output of the function is the DCT transformed image `F`. The function `blockproc` is a MATALB built-in function, which processes an input array `f` in a block-by-block process by applying the function `fun` to each distinct block of size `[L L]` and concatenating the results into the output matrix `F`. The function `fun` is applying the MATLAB built-in 2D DCT function `dct2` to each distinct block `x.data`. It should be noted that we can achieve different processing functions to each distinct block by altering the function `fun`.

Listing 2.3.1: Forward L × L DCT block processing of image array f.

```
function F = blockdct(f,L)
  fun = @(x) dct2(x.data);
  F = blockproc(double(f), [L L], fun);
end
```

Since noise are localized features, therefore, it is most effective for the denoising algorithm to operate within a localized confirmed area. Such operation is known as blocking. Similar remark has been made for pixelwise Wiener filter in Section 2.2. The block size can be varied, with the eight-point DCT is typical in practice, as 8×8 points DCT achieves a balance between computational complexity and overall image quality performance. An example on how to call the function `blockdct` is shown in Listing 2.3.2.

Listing 2.3.2: Forward 8 × 8 DCT block processing of image array f.

```
>> L = 8;
>> F = blockdct(f,L);
```

Similarly, we can create another function to convert the DCT transformed image `F` back into a spatial domain image `g` with the inverse function `idct2` shown in Listing 2.3.3

Listing 2.3.3: Inverse L × L DCT block processing of image array f.

```
function g = blockidct(F,L)
  fun = @(x) idct2(x.data);
  g = blockproc(double(F), [L L], fun);
end
```

An important point in the transform thresholding is about how the threshold is defined. The thresholding can be categorized into hard thresholding, soft thresholding, and combined thresholding. More details on thresholding will be discussed in Section 3.4.1. The choice of which type of thresholding to be applied depends on the image and the noise structures. Because of the high sparsity of the coefficients of the DCT transformed image, hard thresholding are commonly applied in DCT based denoising, which is also the focus of our discussion here. However, it is interesting to take note that the soft [36] and combined [37] thresholding can also be applied in DCT based denoising. Hard thresholding is the simplest thresholding process which sets a transformed coefficient to zero when the magnitude of the coefficient is smaller than a threshold t. The hard threshold function is given by

$$\mathcal{T}(\mathcal{F},t)=\begin{cases}0 & |\mathcal{F}|\leq t,\\ \mathcal{F}, & |\mathcal{F}|>t,\end{cases} \tag{2.14}$$

where $\mathcal{F}$ is the transform coefficients of the image and t is the threshold to be compared with. A MATLAB implementation of the hard threshold function `hthfun(f,th)` is shown in Listing 2.3.4

Listing 2.3.4: Hard thresholding.

```
function g = hthfun(f,th)
  indices = (abs(f)>th);
  g = f.*indices;
end
```

The threshold t can be applied locally in a particular region of the image or applied globally to the overall image, depending on the image structures. To simplify our discussion, we shall confine our discussion on considering a threshold t globally. More details about a locally adaptive t will be discussed in Section 3.4.1.

DCT based denoising can be carried out in non-overlapping, partly overlapping, or fully overlapping blocks. Fully overlap means that the current block and the adjacent block are shifted by only one pixel with respect to each block. However, fully overlap is seldom used in transform domain operation, but it will be more efficient in spatial domain filtering (in terms of both standard and visual quality criteria) [36]. In the following, we shall first consider DCT with non-overlapping 8×8 block. The 8×8 block is chosen to conform to most published literature. Other block sizes might improve filtering performance but only slightly [42]. The case of partly overlap will be discussed in Section 2.3.1.

Consider the AWGN corrupted *Sculpture* image with $\sigma_\eta = 50$ as shown in Figure 1.8(b). The following MATLAB script Listing 2.3.5 performs 8×8 non-overlap blocked DCT hard threshold image denoising. Similar to the importance of the cut-off frequency in ideal filter-based image denoising, the threshold value will affect the performance of the DCT based denoising method. A natural choice is to make $t=\sigma_\eta$ because the noise should pass through the orthogonal transform with the same power. However, we have to account for added image signal power and the HVS masking effects. Therefore, the threshold should be increased to $3\sigma_\eta$ as discussed to remove all transform coefficients that might result in having the noise masked the image signal.

Listing 2.3.5: 8×8 non-overlap blocked DCT hard threshold image denoising.

```
>> sigma=50;
>> th=3*sigma;
>> bsize=8;
>> fdct=blockdct(f,bsize);
>> fdctth=hthfun(fdct,th);
>> g=blockidct(fdctth,bsize);
```

Figure 2.9 Denoised images from a AWGN corrupted *Sculpture* image with σ_η=50 (noisy image: Figure 1.8(b)): (a) by 8×8 non-overlap DCT transform (PSN R=19.7820 dB); (b) by 8×8 half-overlap DCT transform (PSNR=21.0528 dB), both with hard threshold at $t = 3\sigma_\eta = 150$; (c) and (d) are the zoomed-in of the *"book & glass"* of the denoised in (a) and (b), respectively.

The denoised image obtained from Listing 2.3.5 is shown in Figure 2.9(a). The denoised image has PSNR of 19.7820 dB, which is not as high as that from Wiener filtering. Although most of the AWGN has been suppressed, where the edges are sharper. However, blocking effects and texture noise that are signature to non-overlap block transform are observed in the denoised image. If we consider the zoomed-in portion of the *book & glass* (see Figure 2.9(c)), you will find the edges of both the *book & glass* are sharper and well-defined when compared to that of the Wiener filtering denoised image. However, zig-zag like pattern is observed along the sharp edges, which is due to the blocking artifact due to the block based operation of the DCT based denoising. The blocking artifact and checker board problem are more vivid in the homogeneous regions, like on the "*bench*" in the zoomed-in image in Figure 2.9(c). The weak textures are hardly preserved in the denoised image, where the metal meshes behind the *book & glass* are seriously washed out, with the fact that the information of those weak textures are concentrated in the high-frequency band with low energy, which is suppressed by the hard thresholding.

2.3.1 Overlapped Block

To alleviate the blocking artifacts in the denoised image obtained from operating on non-overlapping DCT coefficients, overlapping DCT should be considered. There are a lot of ways to achieve a half-overlap block DCT, and we shall consider the most simple way which separates the half-overlapped block DCT into two transformations. The first transformation will be the same as the normal DCT. While the second transformation will consider an image which is identical to the image in the first transformation but with some of the top and bottom rows removed, and some of the left and right columns removed. The number of rows and columns to be removed will equal to `bsize/2`, half the transform block size. This method is considered not only because we are not interested with the half-overlap block DCT, it is because we would like to demonstrate another property by the end of this section.

Let us consider the denoise performance of the shifted DCT, which can be obtained by the MATLAB script listed in Listing 2.3.6

Listing 2.3.6: Half-shifted block DCT thresholding image denoise.

```
>> sigma=50;
>> th=3*sigma;
>> bsize=8;
>> halfsize=bsize/2;
>> [m,n]=size(f);
>> fshift=f(halfsize+1:m-halfsize,halfsize+1:n-halfsize);
>> fshiftdct=blockdct(fshift,bsize);
>> fshiftdctth=hthfun(fshiftdct,th);
>> gshift=blockidct(fshiftdctth,bsize);
>> g2=double(f);
>> g2(halfsize+1:m-halfsize,halfsize+1:n-halfsize)=gshift;
```

The visual quality of this half block shifted DCT denoised image is almost the same as that obtained in Figure 2.9(a), but a slightly degraded objective quality with PSNR = 19.4952 dB. We can conclude that we have at least learned one thing from Listing 2.3.6, which is the blocked DCT hard threshold denoise method is translation invariant. To construct the half-overlap block denoise result, we have to consider the average between the two images obtained from Listings 2.3.5 and 2.3.6, and the following MATLAB script is created to do the job.

Listing 2.3.7: Half-overlap block DCT thresholding image.

```
>> gx=(g+g2)/2;
>> g=imgtrim(uint8(gx));
```

The denoised image obtained by Listing 2.3.7 is shown in Figure 2.9(b). Most of the blocking noise and the texture noised signature to DCT have been alleviated. The zoomed-in portion of the *book & glass* showing in Figure 2.9(d) has the glass and book having straight edges which does not suffer from blocking noise. Furthermore, most of the background are smooth and the texture in the *Sculpture* image can be observed vividly. The denoised image has an improved PSNR at 21.0528 dB, which is higher than that obtained by the non-overlap block DCT thresholding image denoising, and the half-shifted block non-overlap DCT thresholding image denoising. Checker board problem in the homogenous region and the zig-zag pattern along the sharp edges are also improved in the half-overlap case. The reason why adding these two images together will produce a better image with higher PSNR is because the residue noises in these two denoised images can be considered as two identical but independent Gaussian processes. Therefore, adding them up will have the noises in individual images to be annihilated, and hence a better PSNR. More elaboration on this summation of noisy images with independent but identically noise method to denoise image will be discussed in Chapters 3, 4, and 7. Unfortunately, the washing out of the metal meshes is still severe, which is a common problem in DCT based thresholding denoising.

2.4 Median Filter

An alternative to moving average smoothing and linear transform-based denoising method is the median filter. The median filter replaces the intensity in the noisy image pixel with the median intensity of its neighboring pixels. This is similar to the smoothing filter that replaces the intensity of each pixel with the weighted mean of the intensity of that pixel and it is neighboring pixels, though the median filter is non-linear (and also non-separable in general). In the presence of impulsive noise (or salt and pepper noise), the median filter has been proven to be very effective.

The median filter is a non-linear filter based on ranking the pixel intensities inside a predefined window of the pixel under concern. It is widely used as it is very effective at removing impulsive noise while preserving image edges [8]. The median of the graylevel of the pixels within these window will be the output of the filter, which will replace the original intensity of the pixel situated at the middle of the window. The median is calculated by first sorting all the pixel values from the neighborhood window (mask) $\mathbb{W}$ into numerical order and then replacing the intensity of the pixel being considered with the middle (median) pixel value in this sorted list. This has the effect of forcing pixels with distinct intensities to be more alike with their neighbors, thus eliminating intensity spikes which appear with isolated graylevel. The following equation shows the mathematical operation of a median filter on a pixel located at $[m, n]$.

$$g[m,n] = \text{MED}\{f[m-i, n-j], \quad [i,j] \in \mathbb{W}\}. \tag{2.15}$$

The size of the window $\mathbb{W}$ is mainly $\mathbb{W}_L \times \mathbb{W}_L$ square mask or cross masks, as shown in Figure 2.10, that have some form of symmetry and $\mathbb{W}_L$ is commonly in odd numbers such that there is a well-defined median. However, in some cases, the window size can be even. In such case, the median is computed as the mean of the two pixel values in the middle of the sorted pixel value sequence. The MATLAB script listed in Listing 2.4.1 implements a median filter with square mask for size `w`×`w`. Noted that the mask to be constructed for the boundary pixels in the image will not have all the pixel available to cover the mask. This problem is solved by padding the image in the same way as the boundary being handled for

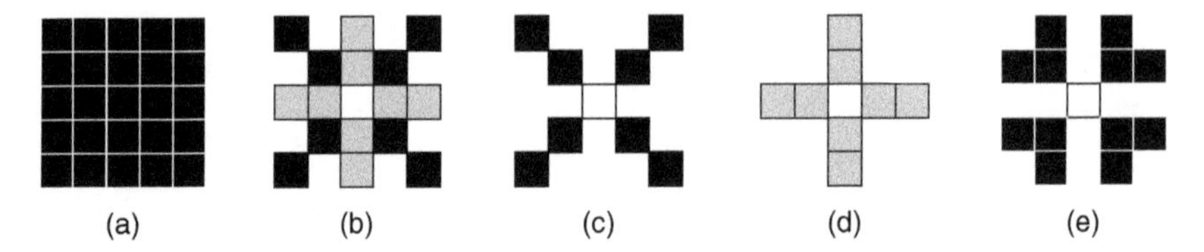

Figure 2.10 Different shapes of the median filter windows (a) 5×5 square, (b) 5×5 neighborhood, (c) 45° neighbors, (d) 90° neighbors, and (e) 60° neighbors.

convolution in linear filter that has been discussed in Section 1.4.1. The function in Listing 2.4.1 uses symmetric padding with size equals to the floor of `w/2`. In addition to padding the image, the median filtered image will also have to be cropped to ditch the padded boundary pixels, such that it will restore the image to the same size as that of the input image.

Listing 2.4.1: Median filter.

```
function g = medianfilter(f,w)

offset = floor(w/2);
fpad = padarray(f,[offset offset], 'symmetric', 'both');

[m,n]=size(fpad);
for i=offset+1:m-offset
    for j=offset+1:n-offset
        block=fpad(i-offset:i+offset,j-offset:j+offset);
        out(i,j)=median(reshape(block,1,[]));
    end;
end;

g = out(offset+1:m-offset,offset+1:n-offset);
end
```

The `medianfilter` function can be applied to filter noisy image, as shown in Listing 2.4.2. The mask size is chosen to be `w=5`, which will conform to all other window sizes used in earlier chapter, and has shown to provide good denoise results in the selected noisy images. The median filtered results for SAP with density 0.05 corrupted *Sculpture* image in Figure 1.10(b), and mixed AWGN with $\sigma_\eta =$ 50 and SAP with density 0.05 corrupted *Sculpture* image in Figure 1.11(a) are shown in Figure 2.11(a) and (b), respectively. The PSNR equals to 21.6335 and 20.1740 dB, respectively. The large difference in the PSNR implies that the performance of the median filter is highly dependent on the nature of the noise.

Listing 2.4.2: Applying median filter.

```
>> w = 5;
>> g = medianfilter(f,w);
```

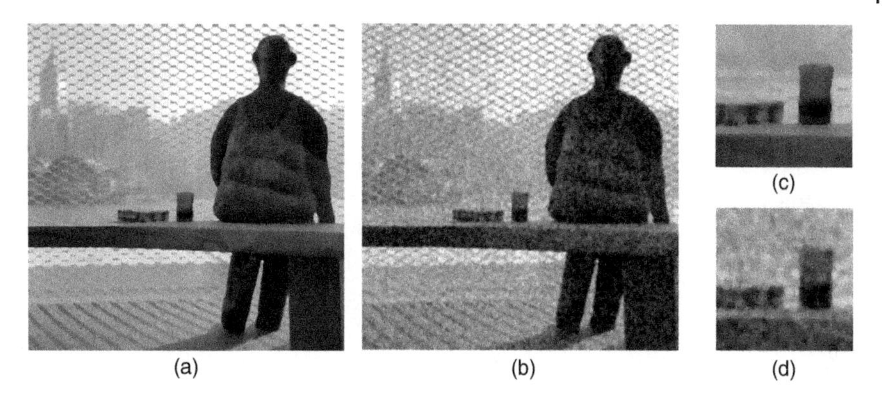

Figure 2.11 Median filter denoised images of (a) SAP corrupted *Sculpture* image with density 0.05 (noisy image: Figure 1.10(b), PSNR: 21.6335 dB), and (b) AWGN and SAP corrupted *Sculpture* image (noisy image: Figure 1.11(a), PSNR: 20.1740 dB) and (c) and (d) the zoomed-in portion of the *book & glass* from (a) and (b), respectively.

The SAP in Figure 1.10(b) and 1.11(a) are observed to have effectively removed in the denoised images by the median filter, as shown in Figure 2.11. However, most of the AWGN in Figure 1.11(a) passes through the median filter retained in the denoised images (see Figure 2.11(b)) because median filter is not a linear lowpass filter. Moreover, not all the SAP can be removed from the noisy image. The SAP around image edges in both SAP corrupted image and the mixed noise corrupted image can still be observed. The SAP in the denoised image around heavy AWGN corrupted region of the mixed noise corrupted image is also retained. The remaining SAP can be more easily observed through the zoomed-in images of the *book & glass* of Figure 2.11(a) and (b), as shown in Figure 2.11(c) and (d), respectively. In Section 4.2.1 will study the noise reduction performance of median filter in order to understand why the SAP does not perform well in particular regions of the image.

2.4.1 Noise Reduction Performance

Because the median filter is a non-linear filter, it is mathematically untractable. An approximated residual noise variance of the median filter denoised image g obtained from an AWGN corrupted image with noise variance σ_η is given by [55]

$$\sigma_{g,\eta}^2 \approx \frac{\sigma_\eta^2}{\mathbb{W}_L + \frac{\pi}{2} - 1} \cdot \frac{\pi}{2}, \tag{2.16}$$

where $\mathbb{W}_L$ is the window size of the median filtering mask. It is vivid that the performance of the median filter depends on two things: (i) the size of the mask and (ii) the distribution of the noise. The mean filtering performance of random noise reduction is better than that of median filtering. But for SAP, and other noises with spatial response of a narrow pulse with a pulse width less than $\mathbb{W}_L/2$, the median filter will provide better performance than that of the mean filter. However, the performance of the median filter highly depends on the window size. When a smaller or larger window size is being applied, there is a high chance that the median filter will blur the denoised image. As a result, the median filtering performance could be improved if the median filtering algorithm can be combined with some kind of moving average filter, or if the window size can adapt to the noise density.

2.4.2 Adaptive Median Filter

Natural image has strong neighboring pixels correlation. The intensity of each pixel is usually close to that of its neighboring pixels, no matter whether it is an edge pixel or a pixel in a smooth region. If the value of a pixel is greater or less than the value in the neighborhood, the pixel is contaminated by the noise. In the reducing-noise process, we sequentially check each pixel, if the value of a pixel is greater than the average value in the mask, then we judge that the pixel is contaminated by the noise and replace it with the median value of the mask; otherwise, we retain the original value of the pixel unchanged. This method not only reduces the computation time but also retains the details of the image as far as possible. The original value of the pixel is replaced with the median value in the mask, and the next process of computation the average value may make full use of the new value of the pixel. This forms an iterative process which will help to decrease the time complexity, and also to improve the noise reduction performance.

The window size determines the effectiveness of the median filter to denoise an image. The smaller the mask, the better the image details are retained, but it also hammers the median filter noise reduction capability, and hence is the best to be used in high SNR environment. The larger the mask, the higher the noise reduction performance, but it also reduces the retention of image details. Therefore, large window size should be applied in low SNR environment. As a result, the median filter window size should adapt to the noise level of the image region under concern. When the window size is small, it is almost sure that there will be no edge, or at most one edge inside the window. Assuming that natural image should be generally smooth in localized region, or localized regions at the two sides of an edge. In that case, the noise level can be determined by inspecting any area that breaks the above rule. The inspection can be carried out by considering the maximum (max) and minimum (min) values of the pixels inside the window under concern. At the same time, the median, maximum, and minimum intensity of pixels inside the window are to be denoted as `Imed`, `Imax`, and `Imin`, respectively. The window size is proper only when the median value of the pixel window under concern is proper, that is it lies in the middle of the pixel value list of all the pixels inside the window. The above test can be mathematically written as

$$Imed - Imin > 0, \quad \& \quad Imed - Imax < 0. \tag{2.17}$$

Otherwise the window is heavily corrupted by noise and thus the median value would not be appropriate to be considered noise free, and hence a larger window size should be used. This logic process can be written as a program loop that increases the window size `w` by 2, i.e. `w=w+2` when Equation 2.17 is not satisfied. The program loop will break when the maximum window size is achieved, or when Equation 2.17 is satisfied. The MATLAB function `medianfilteradp` listed in Listing 2.4.3 implemented this idea for a median filter with adaptive window size.

Listing 2.4.3: Adaptive window median filter.

```
function g = medianfilteradp(f,w,wmax)

fpad = padarray(f,[wmax wmax], 'symmetric', 'both');

[m,n]=size(fpad);
out=zeros(m,n);
finish=0;
for i=wmax+1:m-wmax
    for j=wmax+1:n-wmax
        w1 = w;
        while(finish==0)
            offset = floor(w1/2);
            block=fpad(i-offset:i+offset,j-offset:j+offset);
            Imed = median(reshape(block,1,[]));
            Imin = min(reshape(block,1,[]));
            Imax = max(reshape(block,1,[]));
            Iij = fpad(i,j);
            P1 = Imed - Imin;
            P2 = Imed - Imax;
            if (P1>0 && P2<0)
                A1=Iij-Imin;
                A2=Iij-Imax;
                if(A1>0 && A2<0)
                    finish=1;
                    out(i,j)=Iij;
                else
                    finish=1;
                    out(i,j)=Imed;
                end
            else
                w1 = w1 + 2;
                if (w1>wmax)
                    finish=1;
                    out(i,j)=Iij;
                end
            end
```

```
            end
            finish=0;
        end
    end

    g = out(wmax+1:m-wmax,wmax+1:n-wmax);
    end
```

The application of the function `medianfilteradp` that implements the median filter with adaptive window size is demonstrated by the following MATLAB script Listing 2.4.4, where the adaptive window size will vary between 3×3 and 5×5. The same set of test images is applied and the denoised images are shown in Figure 2.12. Median filter with adaptive window size is very efficient in removing SAP as shown in Figure 2.12(a), not only the visual quality but it has also been reflected in the big improvement on its PSNR, increased to 26.7077 dB, when compared to that of the conventional median filter. However, scattered SAP is still observed in the result image, especially those fallen in the middle of the metal meshes and those along the strong edges. A close up of Figure 2.12(a) at the *Sculpture ear* is shown in Figure 2.12(c) which demonstrates the artifacts. This limitation is due to mismatch between the window size and the feature size. However, this modified median filter has degraded performance on the mixed noise corrupted image, as shown in Figure 2.12(b), where the zoomed-in of the *Sculpture ear* is shown in Figure 2.12(d). Its PSNR has dropped to 17.0446 dB. This is because smaller window was applied, which reduces the associated noise suppression upon the AWGN, that has been discussed in Section 2.4.1. While the SAP in the mixed noise can be effectively suppressed, which is similar to that of the conventional median filter case.

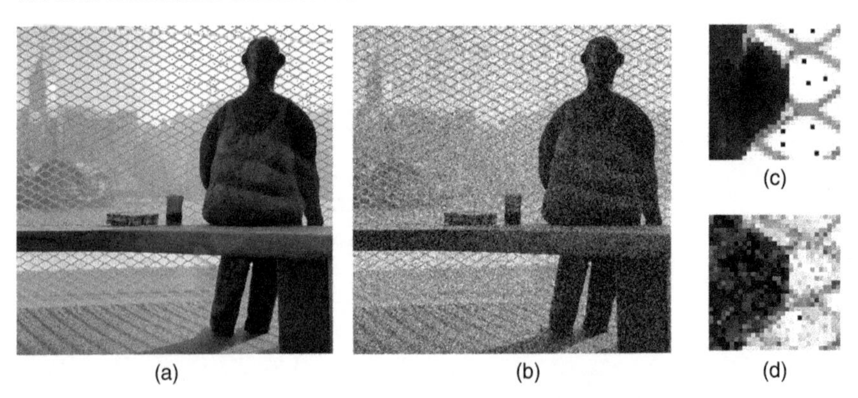

Figure 2.12 Median filter with adaptive window size ranging from 3×3 to 5×5 denoised image: (a) SAP corrupted *Sculpture* image with density 0.05 (noisy image: Figure 1.10(b), PSNR: 26.7077 dB), and (b) AWGN and SAP corrupted *Sculpture* image (noisy image: Figure 1.11(a), PSNR: 17.0446 dB) and (c) and (d) the zoomed-in portion of the *book & glass* from (a) and (b), respectively.

Listing 2.4.4: Applying adaptive window median filter.

```
>> w = 3; wmax = 5;
>> g = medianfilteradp(f,w,wmax);
```

The median filter with adaptive window size has improved the filter performance. To further improve the performance, not only the window size, the window shape will also have to be adaptive, such that the image features can be catered in a better way. Section 2.4.3 will discuss median filter with adaptive window shape.

2.4.3 Median Filter with Predefined Mask

Changing the window shape (mask) helps the median filter to tackle different noise scenarios. To alter the median filter mask is not a difficult task. All we need is to extract the pixels within a predefined mask and then perform median filtering within the extracted pixel vector. The following MATLAB script Listing 2.4.5 implements a median filter to denoise an image `f` using a predefined mask `mask`.

Listing 2.4.5: Median filter with predefined mask.

```
function g = medianfiltermask(f,mask)
% image boundary extension
[m,n]=size(f);
w=size(mask,1);
offset = floor(w/2);
fpad = padarray(f,[offset offset], 'symmetric', 'both');
g=zeros([m,n]);
% median filter
for i=1:m
    for j=1:n
        block=fpad(i:i+w-1,j:j+w-1);
        temp=block(mask~=0);
        g(i,j) = median(temp);
    end
end
end
```

The followings are the MATLAB implementation of the masks line up in Figure 2.10.

```
>>    medsquare = [ 1 1 1 1 1; 1 1 1 1 1; 1 1 1 1 1; 1 1 1 1 1; 1 1 1 1 1];
>>    medneigh  = [ 1 0 1 0 1; 0 1 1 1 0; 1 1 0 1 1; 0 1 1 1 0; 1 0 1 0 1];
>>    med45     = [ 1 0 0 0 1; 0 1 0 1 0; 0 0 1 0 0; 0 1 0 1 0; 1 0 0 0 1];
>>    med90     = [ 0 0 1 0 0; 0 0 1 0 0; 1 1 0 1 1; 0 0 1 0 0; 0 0 1 0 0];
>>    med60     = [ 0 1 0 1 0; 1 1 0 1 1; 0 0 1 0 0; 1 1 0 1 1; 0 1 0 1 0];
```

If we apply medsquare to medianfiltermask, we shall obtain the same median filtered image as that in medianfilter, and that can be implemented in the following MATLAB script

```
>> g = medianfiltermask(f,medsquare);
```

Let us consider the median filter using medneigh applied to both the SAP corrupted image as shown in Figure 1.10(b), and the mixed AWGN with $\sigma_\eta = 50$ and SAP with density 0.05 corrupted image in Figure 1.11(a) by the following MATLAB script.

```
>> g = medianfiltermask(f,medneigh);
```

Compared with that of the conventional median filter with the same window size of 5×5, the use of medneigh mask is able to achieve better suppression of SAP both subjectively and objectively. Showing in Figure 2.13(a) is the result of the SAP image, where the PSNR is 21.8834 dB, which is higher than that of the conventional median filter case. Let's take a closer look of the metal meshes behind the *book & glass*, as shown in Figure 2.13(c), it can be observed that the weak lines are better preserved, and more SAP sitting on the strong edges are suppressed. A similar subjective performance is obtained for the mixed noise corrupted image, as shown in Figure 2.13(b) and the corresponding zoomed-in image, as shown in Figure 2.13(d). Nonetheless, its PSNR is degraded when compared to that of the conventional median filter. Now we have learned that the denoised image performance depends on which mask we use in the median filter. Can we do better than a wild guess to use any particular mask? One of the ways to answer this question is in Section 2.4.4, where median of median is applied.

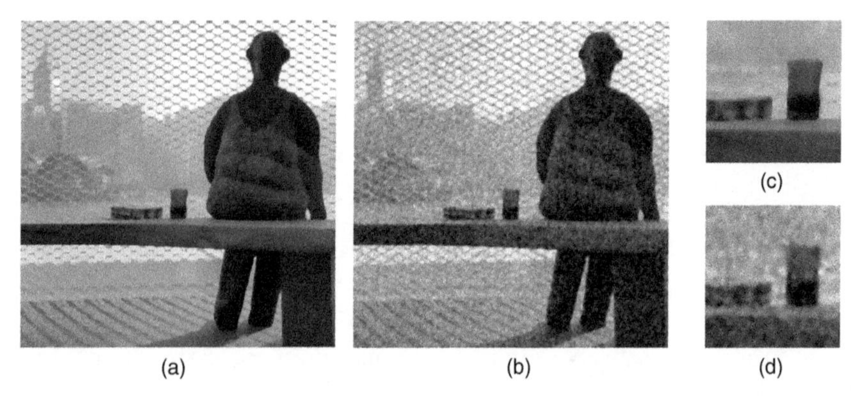

Figure 2.13 Median filter with mask medneigh denoised image: (a) SAP corrupted *Sculpture* image with density 0.05 (noisy image: Figure 1.10(b), PSNR: 21.8834 dB), (b) AWGN and SAP corrupted *Sculpture* image (noisy image: Figure 1.11(a), PSNR: 19.7594 dB), and (c) and (d) the zoomed-in portion of the *book* & *glass* from (a) and (b), respectively.

2.4.4 Median of Median

Instead of considering the median of only one window shape, the output of median of different window shapes, such as those listed in Figure 2.10 can be considered at the same time to take the geometric shape advantage to achieve better denoising. If the mask set in Figure 2.10 is considered, this mask set will enhance the detection of image edge with 45°, 60°, and 90°, and hence should be able to preserve image edge better in the denoised image. The median of all the outputs obtained by different median window will be the intensity of the pixel under concern [47].

Listing 2.4.6: Median of median filter.

```
function g = medianmedian(f)
medsquare = [ 1 1 1 1 1; 1 1 1 1 1; 1 1 1 1 1; 1 1 1 1 1; 1 1 1 1 1];
medneigh  = [ 1 0 1 0 1; 0 1 1 1 0; 1 1 0 1 1; 0 1 1 1 0; 1 0 1 0 1];
med45     = [ 1 0 0 0 1; 0 1 0 1 0; 0 0 1 0 0; 0 1 0 1 0; 1 0 0 0 1];
med90     = [ 0 0 1 0 0; 0 0 1 0 0; 1 1 0 1 1; 0 0 1 0 0; 0 0 1 0 0];
med60     = [ 0 1 0 1 0; 1 1 0 1 1; 0 0 1 0 0; 1 1 0 1 1; 0 1 0 1 0];
% image boundary extension
[m,n]=size(f);
w=5; % all the above mask are defined with a 5x5 window
offset = floor(w/2);
fpad = padarray(f,[offset offset], 'symmetric', 'both');
g=zeros([m,n]);
t=zeros(5,1);
% median filter
for i=1:m
    for j=1:n
        block=fpad(i:i+w-1,j:j+w-1);
        t(1)=median(block(medsquare~=0));
        t(2)=median(block(medneigh~=0));
        t(3)=median(block(med45~=0));
        t(4)=median(block(med90~=0));
        t(5)=median(block(med60~=0));
        g(i,j) = median(t);
    end
end
end
```

Let us consider the median of median filter using `medianmedian` applied to both the SAP corrupted image in Figure 1.10, and the mixed AWGN with $\sigma_\eta = 50$ and SAP with density 0.05 corrupted image in Figure 1.11(a) by the following MATLAB script.

```
>> g = medianmedian(f);
```

Figure 2.14(a) shows the denoised result of the SAP corrupted image and the corresponding zoomed-in image of the *book & glass* is shown in Figure 2.14. It can be observed that the SAP suppression is further improved when compared to that of the median filter using `medneigh` mask. The PSNR is also improved and

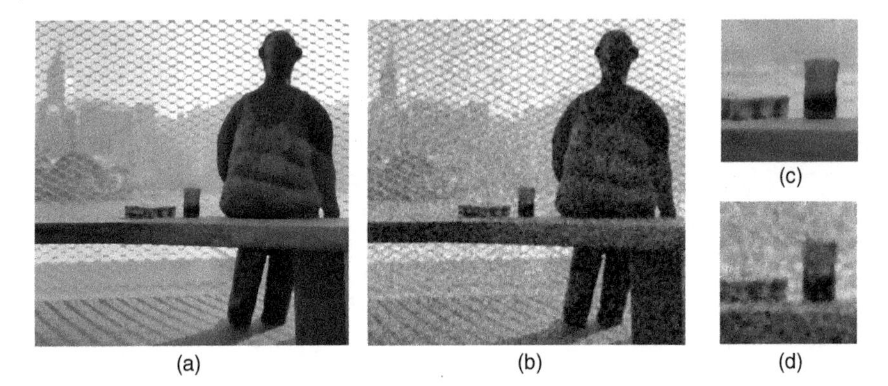

Figure 2.14 Median of median filter denoised image: (a) SAP corrupted *Sculpture* image with density 0.05 (noisy image: Figure 1.10(b), PSNR: 21.9932 dB), (b) AWGN and SAP corrupted *Sculpture* image (noisy image: Figure 1.11(a), PSNR: 20.2311 dB), (c) and (d) the zoomed-in portion of the *book* & *glass* from (a) and (b), respectively.

increased to 21.9932 dB when the median of median filter is used. Improvement in both visual quality and objective performance is also observed in the mixed noise corrupted images, when compared to that of the median filter using `medneigh` mask, where the PSNR is 20.2311 dB. The denoised image is shown in Figure 2.14(b) and the corresponding zoomed-in image is shown in Figure 2.14(d). Reader should have already find out the median of median method is another way to combine a number of denoised image together to obtain a better denoised image. Previously we have discussed the averaging of multiple denoised images to obtain a better denoised image. Actually, there are a lot of different methods to combine denoised images to obtain a better denoised image, and that will be the topic in Chapters 3, 4, and 7.

2.5 Summary

The need for image enhancement and restoration is encountered in many practical applications. For instance, distortion due to additive white Gaussian noise (AWGN) can be caused by poor quality image acquisition, images observed in a noisy environment or noise inherent in communication channels. Linear filtering and smoothing operations have been widely used for image restoration because of their relative simplicity. However, since these methods are based upon the assumption that the image signal is stationary and formed through a linear system, their effectiveness is generally acceptable but limited. In reality, real-world images have typically non-stationary statistical characteristics. They are formed through a nonlinear system process where the intensity distribution arriving at the imaging system is the product of the reflectance of the object or the scene of interest and the illumination distribution falling on the scene. There also exists various adaptive and nonlinear image restoration methods that account for the variations in the local statistical characteristics. These methods achieve better enhancement and restoration of the image while preserving high-frequency features of the original image such as edges.

Linear filtering techniques are extensively used in image denoising. They are very popular because of the mathematical simplicity, backed by the linear system theory and computational efficiency. In spite of these, not all denoising problems can be satisfactorily addressed through the use of linear filters. Particularly, linear filter performs poorly in the case of impulse noise. They will blur edges and ruin lines, and other fine image textures. The linear filtering is a kind of Fourier method, and the Fourier method has been extended in the past 30 years to other linear-space-frequency transforms such as the blocked DCT [53] and wavelet transform (will be as detailed in Chapter 3).

The linear filtering method can also be extended to nonlinear filtering techniques. Among all nonlinear filtering techniques, the median filter is the most well-known nonlinear filter. The median filter has found to be very effective in suppressing salt and pepper noise. We have also discussed varies improvement to median filtering. In particular, we have discussed the median of median filtering method that combines multiple median filter denoised image obtained with different median masks which generates the best denoised image among all the median mask filtered denoised image.

Removing noise from the original signal is still a challenging problem for researchers. Each image denoising algorithm has its assumptions, advantages, and limitations. Chapters 3–7 will present various image denoising methods. Image denoising remains challenging and interesting, and we hope the readers will start loving these interesting research topics after reading this book.

Exercises

2.1 Compare the effect of different passband roll off factor σ of the Gaussian filter with kernel size of 5×5 in the denoising of the AWGN corrupted *Sculpture* image (source image: Figure 1.8(b)) and the AWGN and SAP mixed corrupted *Sculpture* image (source image: Figure 1.11(a)).

2.2 Compare the effect of different filter kernel sizes of the Wiener filter in the denoising of the AWGN corrupted *Sculpture* images with different noise variances (source image: Figure 1.8(a) and (b)) and the AWGN and SAP mixed corrupted *Sculpture* image (source image: Figure 1.11(a)).

2.3 Implementation of pixel-based Wiener filter.

1. The following is pseudo-code illustrating the pixel-based Wiener filter. Implement it in MATLAB

```
Pad the image boundary with width equals to the
    selected window size
For i=1 to column size
        For j=1 to row size
                u = compute mean for local window
                    around (i,j);
                (remember the pixel location offset)
                s2 = compute standard deviation for
                     local window around (i,j);
                (remember the offset)
                g(i,j) = u+((s2-noise)/s2)*(padded f(i
                   ,j) - u);
                (remember the offset)
        end;
end;
```

2. Execute the developed function with window size 3×3 and 5×5 and compare your results with that obtained from MATLAB built-in function `wiener2`.

2.4 Denoise performance versus transform block size: Consider the denoising of the noisy image f of the *Sculpture* image corrupted by AWGN with $\sigma_\eta = 25$.

1. Perform the block-based DCT hard threshold denoising on f with block size 4×4, 8×8 and 16×16. Report the PSNR of each denoised image, and state the visual quality of each denoised images. What conclusion can be drawn from this experiment?
2. Generate a new denoised image from averaging the three denoised images obtained in exercise 2.4.1. Please report the PSNR of the denoised image and compare the visual quality of the denoised image with those obtained in exercise 2.4.1.
3. Suggest a method to improve the visual quality of the denoised image obtained in exercise 2.4.2.

2.5 Develop a MATLAB function that accepts the following inputs.

1. f : noisy input image.
2. *wsize* : patch size (odd number).
3. ϵ: constant to compare with variance in a patch.

The output of the function is g which is initiated by f. The function will process every patch of size *wsize* $\times$ *wsize* in the noisy image f, such that when the absolute difference of the maximum or minimum pixel intensity within the patch with the mean pixel intensity within the patch is larger than ϵ, the intensity of the corresponding center pixel of the patch in the output image will be replaced by the median intensity of the patch. Otherwise, the intensity of the corresponding center pixel of the patch of the output image will be replaced by the mean intensity of the patch. The function will output the modified image g.

Perform experiments on your developed function with the noisy image *Sculpture* image corrupted by AWGN with $\sigma_\eta = 25$ and SAP with density 0.05 with *wsize* = 3 and varying ϵ between 40 and 200 with step size of 1. Plot the PSNR versus ϵ. Comment on the denoise performance of the output image on the varying ϵ. Explain what do you observe. What happens when you change *wsize*?

3

Wavelet

Denoising by thresholding on the coefficients obtained from projecting the image onto transform basis functions has been discussed in Chapter 2. However, these denoising methods have limitations, which are caused by the global nature of the basis functions in the spatial domain, thus, the thresholding operation will affect the whole image. In other words, even if the noise is localized in a particular spatial location, the thresholding cannot take advantage of this property to produce a better denoised image. The localization property has been exploited by blocked transform in Section 2.3, where the block transformation is equivalent to the application of rectangular spatial window to extract a finite length sub-image in certain location before the transformation, and hence localized the transformation and thus the threshold effect. However, the discontinuity between blocks will create blocking artifacts in the denoised image, and hence a better localization technique other than rectangular windowing is required. Block transformation with overlapped blocks has been discussed in Section 2.3.1 to alleviate the blocking artifacts. When the transformation between blocks achieves maximum overlap, and all the transform basis functions satisfy a list of regularity requirements, the transformation is known as the wavelet transform. The wavelet transform provides multiresolution analysis which has very good signal localization property.

Readers interested in wavelet transform can learn about its' fundamental theory from [48]. This chapter will only discuss the application of wavelet transform to image denoising with the assumption that the readers have learned wavelet theory from other classes. Having said that, we shall still introduce wavelet transform in Section 3.1 for the virtue of introducing the equations, symbols, and MATLAB scripts used in this book.

Digital Image Denoising in MATLAB, First Edition. Chi-Wah Kok and Wing-Shan Tam.

Companion website: www.wiley.com/go/kokDeNoise

3.1 2D Wavelet Transform

The 2D Discrete Wavelet Transform (DWT) on an image f can be computed by first performing 1D DWT (horizontally) on the rows with lowpass subband filter $h_0[n]$ ($H_0(z)$ in z-transform) and highpass subband filter $h_1[n]$ ($H_1(z)$ in z-transform), respectively. Then we perform the same 1D DWT on the columns (vertically) with the same set of subband filters for both the lowpass and highpass subband signals obtained in the horizontal analysis, as shown in Figure 3.1 taking the *Sculpture* image as an example. In this book, we adopt the dyadic ratio as the wavelet decomposition and reconstruction sampling rates (as denoted by "$\uparrow 2$" and "$\downarrow 2$" in Figure 3.1). As a result, there will be four subband images f^1_{LL}, f^1_{LH}, f^1_{HL}, and f^1_{HH}, where the L and H indices are used for the lowpass and highpass filtering in 1D DWT, respectively, and the superscript "1" refers to 1-level wavelet decomposition. The following MATLAB script, Listing 3.1.1, makes use of the MATLAB wavelet library function `dwt2` to perform a single-level two-dimensional wavelet decomposition on the image array `f`. Different wavelet functions are available with the `dwt2` function, and in Listing 3.1.1, the wavelet `db1`, the Daubechies family wavelet with order 1 has adopted, which is the Haar wavelet.

Listing 3.1.1: Single-level discrete Haar wavelet transform.

```
[fll1,flh1,fhl1,fhh1]=dwt2(f,'db1','per');
```

The `dwt2` computes the approximation coefficients matrix `fll1` (f^1_{LL}), and details coefficients matrices `flh1` (f^1_{LH}), `fhl1` (f^1_{HL}), and `fhh1` (f^1_{HH}), (horizontal, vertical, and diagonal, respectively), which are shown in Figure 3.1(c). The "1"

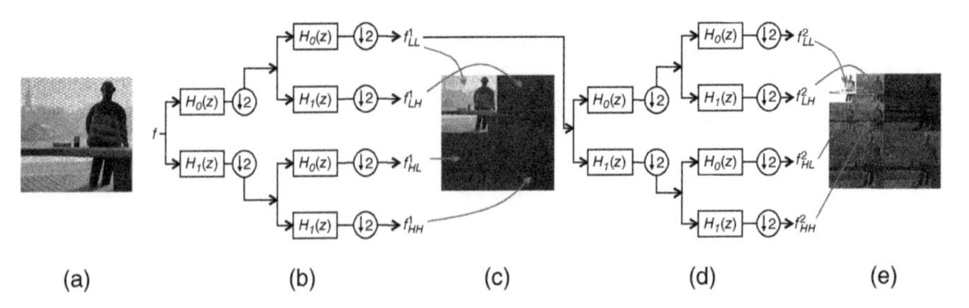

Figure 3.1 Illustration of 2D wavelet decomposition of the *Sculpture* image. (a) Filter bank structure for 2D wavelet image transformation, (b) the corresponding first-level wavelet decomposition, and (c) the subband images stitched together to show the invariance of the overall spatial size, (d) second-level dyadic decomposition, with (e) the subband images from first and second level decomposition stitched together (the subband images looks darker because brightness scaling has been applied to squeeze the dynamic range of all subband images to be displayed in the same figure).

at the last index of each wavelet coefficients indicates this is the 1-level wavelet decomposition. Filtering is a signal expanding operation. The subband image size equals to `floor((size(f)+size(lf)-1)/2)`, where the filtered signal has a length expansion of `lf`, which is the length of the subband filter. At the same time, the total signal length will be divided by 2 because of the subsampling. The floor operation is to ensure the subband signal length after subsampling is an integer. Because the subband filters in `db1` have the same length, therefore, the subband signal length will be of `size(fll1)` = `size(flh1)` = `size(fhl1)` = `size(fhh1)`. Finally, the third parameter `per` in Listing 3.1.1 is the extension mode parameter which is set to periodization to extend the signal along its boundaries. In other words, the filtered signal will also be a periodic signal, where all the signal outside `ceil(size(f)/2)` can be removed without affecting the information content. As a result, all the subband signal will have the same length and equals `ceil(size(f)/2)`.

A 2-level decomposition can be obtained by replacing `f` with `fll1` (f^1_{LL}) in Listing 3.1.1, as shown in Listing 3.1.2. The subband images obtained by 2-level wavelet decomposition are f^2_{LL}, f^2_{LH}, f^2_{HL}, f^2_{HH}, f^1_{LH}, f^1_{HL}, and f^1_{HH}, where the superscript “1” and “2” are used to separate the subband images obtained at different levels of wavelet decomposition. The subband images are shown in Figure 3.1(e). Since the variables in MATLAB are not allowed to use superscript, therefore, the variables for each subband images are appended with the number “1” or “2” in Listing 3.1.2.

Listing 3.1.2: Two-level discrete Haar wavelet transform.

```
[fll1,flh1,fhl1,fhh1]=dwt2(f,'db1','per');
[fll2,flh2,fhl2,fhh2]=dwt2(fll1,'db1','per');
```

After obtaining the subband images using Listing 3.1.2, Figure 3.1(c) and (e) can be obtained by Listing 3.1.3. Please noted that to avoid subband image size mismatch due to the signal expansion problem of the filtering operation, the composite `level2` subband image is resized by the built-in MATLAB function `imresize` to ensure its size is compatible to that of the subband images in level 1 decomposition.

Listing 3.1.3: Wavelet image display.

```
>> level2 = [brightnorm(f,fll2) brightnorm(f,flh2);
    brightnorm(f,fhl2) brightnorm(f,fhh2)];
>> level1 = [brightnorm(f,imresize(level2,size(fll1))),
    brightnorm(f,flh1); brightnorm(f,fhl1) brightnorm(f,fhh1)];
>> figure; imshow(imgtrim(unit8(level1)));
```

The spatial image can be reconstructed from the subband images using the MATLAB built-in function `idwt2`, which is the inverse wavelet transform function. An example of the reconstruction of image `g` from the 2-level DWT coefficients `fll2`, `flh2`, `fhl2`, `fhh2`, `flh1`, `fhl1`, and `fhh1` is shown in Listing 3.1.4. The reader should note that the same boundary extension method should be used in the forward and inverse wavelet transform to avoid introducing any noise into the reconstructed image.

Listing 3.1.4: Inverse discrete wavelet transform.

```
fll1=idwt2(fll2,flh2,fhl2,fhh2,'db1','per');
g=idwt2(fll1,flh1,fhl1,fhh1,'db1','per');
```

Higher level wavelet decomposition can be obtained similar to obtaining level 2 wavelet decomposition from the result of level 1 wavelet decomposition. A lot of things can be done with the high-level wavelet decomposition. In this book, we are interested in at least three applications: (i) noise power estimation, (ii) image denoising, and (iii) image fusion. The first two applications will be discussed in this chapter, while the third application will be discussed in this chapter and also in Chapter 7.

3.2 Noise Estimation

Following the same argument as that presented in Chapter 2 for Figure 2.2, the high-frequency component of a noisy image is almost completely occupied by noise. Since on each level of the wavelet decomposition, it will divide the image into lowpass (detail) subband image, and highpass (approximate) subband image. The highpass subband image will contain mostly noise component. When we increase the number of wavelet decomposition levels to J, the spectral bandwidth of f^J_{HH} is so narrow (compared to that of the spectral domain of the original image) that it is almost sure it will only contain noise from the image f. This assumption can also be reached from the sparse representation of image signal in the wavelet domain. Since the f^J_{HH} subband only contains noise, if we further assume the noise is independent Gaussian random variable with zero mean and variance $\hat{\sigma}_\eta$ at each level of wavelet decomposition, where $\hat{\sigma}_n$ can be estimated via median absolute deviation method as discussed in Equation 1.14. As a result, for a J-level wavelet decomposition,

$$\hat{\sigma}_\eta = \frac{MAD(|f^J_{HH}|)}{0.6745}. \tag{3.1}$$

The constant 0.6745 is also adopted in [18] in the original discussion of wavelet denoising by thresholding. We can make use of the built-in MATLAB function mad to estimate the noise $\hat{\sigma}_\eta$ from the subband image f_{HH}^J, as shown in MATLAB Listing 3.2.1

Listing 3.2.1: Wavelet domain image noise estimation.

```
function sigma = waveletnoiseest(fhhJ)
  sigma = mad(abs(fhhJ(:)),1)/0.6745;
end
```

Although the original image is responsible for a few large amplitude outliers, these few coefficients have little impact on the median operation. Figure 3.2 illustrates the distribution of the f_{HH}^3 subband image coefficients obtained by the MATLAB Listing 3.2.2.

Listing 3.2.2: Plotting the histogram of fhh3 wavelet coefficients.

```
>> histogram(fhh3);
```

Note that this distribution is highly symmetric and has zero mean which resembles a Gaussian distribution and thus empirically proves the above claim. Consider the *Sculpture* image corrupted with AWGN having $\sigma_\eta = 25$. Table 3.1

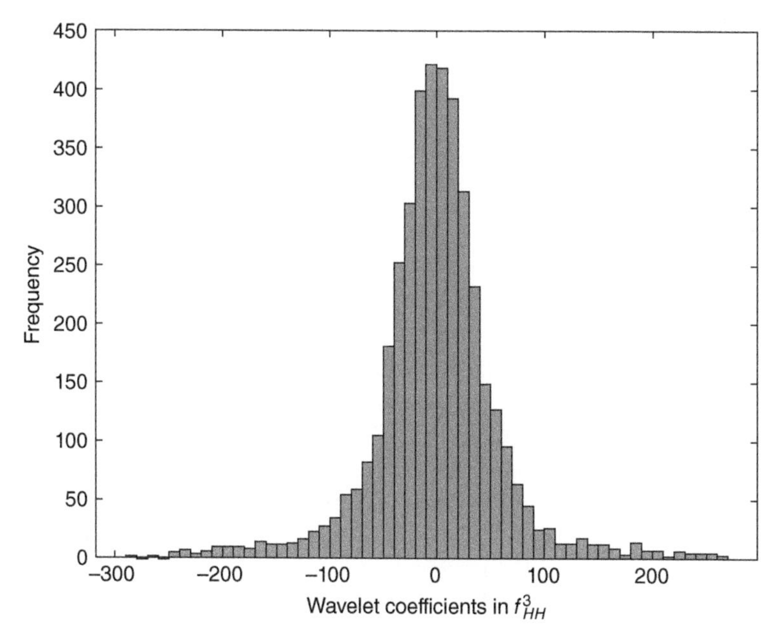

Figure 3.2 The histogram of f_{HH}^3 subband.

Table 3.1 Statistical properties of the subband signals of AWGN corrupted *Sculpture* image with $\sigma_\eta = 25$ with 3-level of wavelet decomposition ($J = 3$).

Size(HH3)	Mean(HH3)	Variance(HH3)	Std(HH3)	MAD(|HH3|)	$\hat{\sigma}_\eta = \frac{\text{Median}(\|\text{HH3}\|)}{0.6745}$
128×128	-0.7166	3.5439×10^3	59.5306	16.7500	24.8332

lists the statistics as well as the estimated noise standard deviation $\hat{\sigma}_\eta$ obtained by Equation 3.1. It is vivid that the wavelet image noise estimation method in Equation 3.1 with 3-level of decomposition, $J = 3$, has provided a very accurate estimate, with $\hat{\sigma}_\eta = 24.8332$ where the true noise standard deviation is $\sigma_\eta = 25$. The rest of the column in Table 3.1 are obtained by the following MATLAB script.

```
>> meanHH = mean(mean(fhh3(:)));
>> varianceHH = var(reshape(fhh3(:,:),[],1));
>> stdHH = sqrt(varianceHH);
>> madHH=mad(abs(fhh3(:)),1);
>> sigmaHH = madHH/0.6745;
```

The wavelet based method yields the best estimate of the noise variance, as compared to the noise estimation method presented in Chapter 1. This is because the discrete wavelet transform performs a significant degree of localization both spatially and in frequency domain. Consequently, the f_{HH}^J subband of a J-level wavelet decomposition, has most of the wavelet coefficients due to noise. Although, the noise-free image has contributed a few large amplitude coefficients in f_{HH}^J, these few coefficients have little impact on the median operator, and hence an accurate noise variance estimation can be obtained.

3.3 Wavelet Denoise

Now we know wavelet can do a decent job in estimating noise variance. How about applying wavelet to denoise images? Let us first consider the nature of the wavelet transform, where the fine resolution level will only contain detail information and that the main features of the image are captured in the low-resolution levels. Informally, we can say that the image are being packed into a few wavelet coefficients. On the other hand, the AWGN is invariant with respect to the wavelet transform, and passes to the wavelet domain (the fine scale) structurally unaffected. Therefore, performing wavelet transform on the noisy image, deleting the finest resolution levels, and back-transforming will give a denoised image. The modified subband images will have a smoothing effect on the reconstructed image and will reduce some of the noise. This method resembles the filtering method discussed in Chapter 2 with the Fourier basis being replaced by the wavelet basis, and the cutoff frequency of the lowpass subband filter forms the hard threshold in the wavelet domain.

Example 3.1 This example investigates how to realize a threshold method in wavelet domain that can achieve better denoise performance than that of Fourier based hard thresholding image denoising techniques. Consider a 1D signal that is extracted from column 400 of the *Sculpture* image, which is plotted in Figure 3.3, where Figure 3.3(a) is obtained from the noise-free image. This 1D signal and can be decomposed by 1D wavelet transform using `dwt` in the MATLAB wavelet library, as shown in Listing 3.3.1.

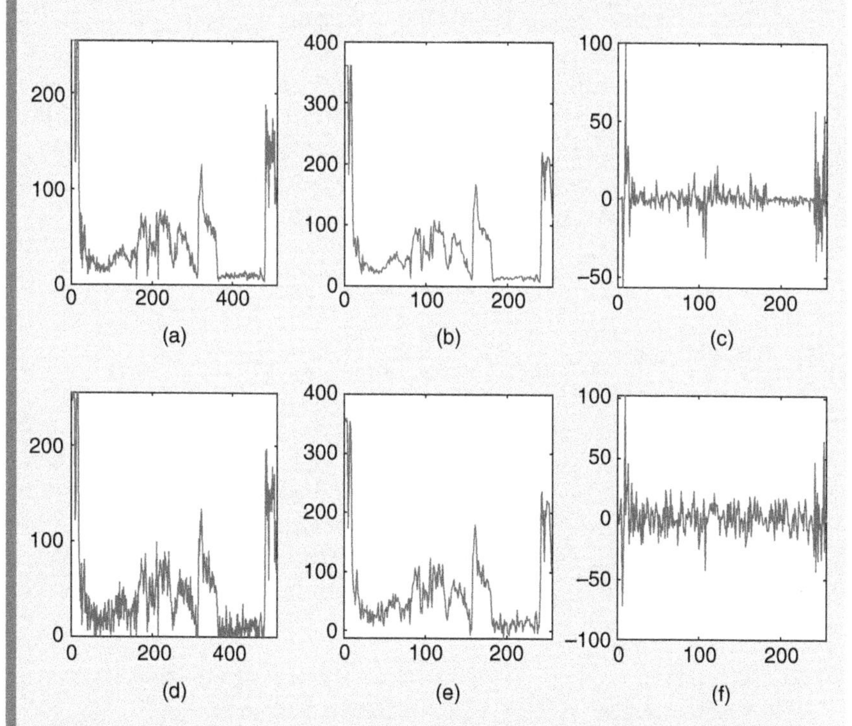

Figure 3.3 Column 400 of the *Sculpture* image: (a) from the noise-free image, (b) from the lowpass subband image of the noise-free image, (c) from the highpass subband image of the noise-free image, (d) from the AWGN corrupted image with $\sigma_\eta = 10$ (noisy image: Figure 1.8(a)), (e) from the lowpass subband image of AWGN corrupted image with $\sigma_\eta = 10$ (noisy image: Figure 1.8(a)), and (f) from the highpass subband image of AWGN corrupted image with $\sigma_\eta = 10$ (noisy image: Figure 1.8(a)).

Listing 3.3.1: 1D discrete wavelet transform of selected column in *Sculpture* image.

```
>> fcol = f(:,400);        % extracting column 400 from image f
>> figure;subplot(1,3,1),plot(fcol),xlim([0,512]),ylim([0,255]);
>> [fl, fh] = dwt(fcol,'db1','per');
>> subplot(1,3,2),plot(fl),xlim([0,256]);
>> subplot(1,3,3),plot(fh),xlim([0,256]);
```

The lowpass and highpass subband signals of the extracted column from the noise-free image are plotted in Figure 3.3(b) and (c), respectively. Similarly, Figure 3.3(d) is obtained from the same column of the AWGN corrupted image with $\sigma_\eta = 10$, where the extracted column of the noisy image is shown in Figure 1.8(a). By replacing f in Listing 3.3.1 with that extracted from the noisy image, we shall obtain the subband signals, as plotted in Figure 3.3(e) and (f), respectively. It can be observed by comparing Figure 3.3(c) and (f) that there are only a few large coefficients in the detail signal, and that the AWGN is transformed into the wavelet coefficients, uniformly at all scales. Therefore, it can be effectively removed by setting all wavelets detail coefficients with magnitude smaller than a predetermined threshold to zero, thus leading to the wavelet thresholding denoise method to be discussed in Section 3.3.

3.4 Thresholding

The example in Figure 3.3 empirically confirmed that the detail coefficients with small magnitude are dominated by noise. Replacing coefficients with magnitude below a certain *threshold* value by some predetermined values, such as zero, will help to suppress the noise in the image reconstructed from the inverse wavelet transform of the modified wavelet coefficients. This is known as the wavelet thresholding denoise method, which is shown to be effective because of the following assumptions

- The decorrelating property of the wavelet transform creates a sparse signal, where a larger number of coefficients are zero or close to zero.
- The noise property preserves across all coefficients in all wavelet levels.
- A threshold value that separates the image signal from the noise can be determined from the noise level.

Based on the above assumptions, the actual thresholding operation will have to be divided into three consecutive processes: (i) performing wavelet decomposition of a noisy image f with a chosen level J; (ii) finding the right threshold value t and applying it to the threshold function $\mathcal{T}$ to the noisy wavelet coefficient $\mathcal{F}$, where the output of the function is the denoised signal $\mathcal{G}$; and (iii) performing inverse wavelet decomposition on the denoised signal $\mathcal{G}$ to generate the denoised image g in spatial domain. A block diagram of this process is given in Figure 3.4, which is implemented in Listing 3.4.1.

Listing 3.4.1: Wavelet thresholding denoising algorithm.

```
function [g j th] = waveletth(f, fsize, j, level)
  th = 0;
  [fll, flh, fhl, fhh] = dwt2(f,'db1','per');
  j=j+1;
  if j<level [fll j th] = waveletth(fll, fsize, j, level);
  else th = thest(fhh,fsize);
  end
  flh = thfun(flh, th); fhl = thfun(fhl, th); fhh = thfun(fhh, th);
  g = idwt2(fll, flh, fhl, fhh, 'db1', 'per');
  j=j-1;
end
```

The function `waveletth` is a recursive function, where it recursively calls itself to perform wavelet decomposition at the detail subband image f_{LL} until it reaches the input level of decomposition `level`. The thresholding function is being implemented with `thfun` (i.e. $\mathcal{T}$ in Figure 3.4). The thresholding value `th` (i.e. t in Figure 3.4) is estimated by the function `thest` at the highest level of the decomposed wavelet detail subband image. For the purpose of demonstration in this book, the '`db1`' wavelet and periodic extension will be chosen to implement MATLAB Listing 3.4.1.

The functions used in `waveletth` include `thest`, threshold value estimation function, which will be invoked on the highest scale of the wavelet coefficients. The theory and MATLAB implementation of `thest` will be given in a sequel. The other function `thfun` is the wavelet coefficients threshold function which makes use of the thresholding coefficient `th` obtained from `thest` to threshold all levels of wavelet details coefficients. Section 3.4.1 will discuss a number of commonly applied and also very efficient threshold functions.

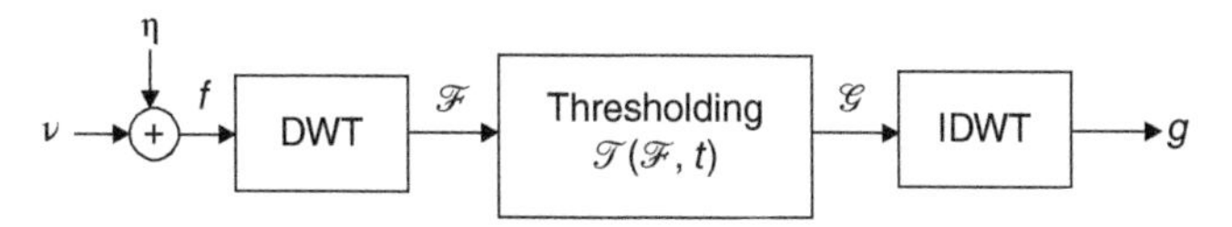

Figure 3.4 The three steps involved in the wavelet denoising process.

3.4.1 Threshold Function

When wavelet thresholding is first presented by Donoho [17], he also introduced the hard and soft thresholding. We have applied hard thresholding in Chapter 2 under transform domains, such as ideal filter denoising in Section 2.1, and DCT based denoising in Section 2.3. A proper definition of the hard thresholding function with threshold t is given by

$$\mathcal{T}(\mathcal{F}, t) = \begin{cases} 0 & |\mathcal{F}| \le t, \\ \mathcal{F}, & |\mathcal{F}| > t, \end{cases} \tag{3.2}$$

where $\mathcal{F}$ is the wavelet transformed detail coefficients (f^k_{LH}, f^k_{HL}, and f^k_{HH} at level k). A MATLAB implementation of the hard threshold function `hthfun(f,th)` that can be applied to replace `thfun` in Listing 3.4.1 is shown in Listing 3.4.2

Listing 3.4.2: Hard thresholding.

```
function g = hthfun(f,th)
  indices = (abs(f)>th);
  g = f.*indices;
end
```

The transfer function of the hard threshold function is plotted in Figure 3.5 with a dotted line. It is vivid that the hard threshold function is a "keep or kill" procedure. It does not affect the wavelet coefficients with magnitude greater than

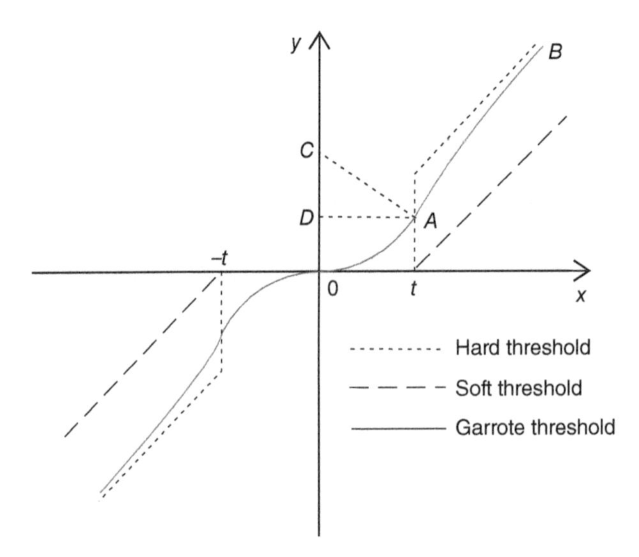

Figure 3.5 Threshold functions $y = \mathcal{T}(x, t)$.

the threshold level t. However, the hard thresholding method will introduce discontinuity in the wavelet coefficients, which might yields abrupt artifacts in the denoised signal, especially when the noise level is significant [11].

The signal discontinuity problem can be alleviated by the soft thresholding operator which is defined as

$$\mathcal{T}(\mathcal{F}, t) = \text{sign}(\mathcal{F}) \max(0, |\mathcal{F}| - t), \tag{3.3}$$

Comparing the transfer function of the soft threshold function plotted in Figure 3.5 with a dashed line and that of the hard threshold function, it can be observed that the soft thresholding function shrinks coefficients above the threshold in absolute value. Although all wavelet coefficients will get shrink which is not intuitively appealing and seems unnatural. The denoised image can still be faithfully reconstructed. The continuity of the soft thresholding has some advantages over the discontinuous hard threshold function. The continuity of the soft thresholding function makes the denoising algorithm mathematically tractable. Moreover, hard thresholding does not even work with some sophisticated denoising algorithms [30]. Furthermore, pure noise coefficients may pass through the hard threshold and appear as annoying "blips" in the output. On contrary, the soft thresholding shrinks these false structures. However, for some class of signals, the hard threshold results are found to be better than that obtained by soft thresholding, despite the discontinuity problem.

A MATLAB implementation of the soft threshold function `sthfun(f,th)` that can replace `thfun` in Listing 3.4.1 is shown in Listing 3.4.3.

Listing 3.4.3: Soft thresholding.

```
function g = sthfun(f,th)
  indices = (abs(f)>th);
  sgnf = f>0;
  g = f.*indices+th*(indices.*(1-2*sgnf));
end
```

The discontinuity and wavelet coefficient shrinkage dilemma of the hard and soft threshold function has led to the development of threshold function that modifies the hard threshold function to create a continuous threshold function. Such that the wavelet coefficients obtained from this new threshold function will reduce the signal discrepancy between the noise-free signal and the denoised signal. One of such smooth threshold functions is the Garrote threshold function [23, 24].

$$\mathcal{T}(\mathcal{F}, t) = \begin{cases} 0 & |\mathcal{F}| \le t, \\ \mathcal{F} - \frac{t^2}{\mathcal{F}}, & |\mathcal{F}| > t. \end{cases} \tag{3.4}$$

The transfer function in Equation 3.4 has been plotted in Figure 3.5 with a solid line together with that from the hard and soft threshold functions having the same threshold t. A MATLAB implementation of the Garrote threshold function `gthfun` can be used to replace `thfun(f,th)` in Listing 3.4.1 is listed in Listing 3.4.4

Listing 3.4.4: Garrote thresholding.

```
function g = gthfun(f,th)
  indices = (abs(f)>th);
  t = f.*indices;
  nonzerot = t+(1-indices);
  g = (nonzerot - th^2./nonzerot).*indices;
end
```

The Garrote threshold function helps to alleviate the discontinuity problem. However, due to the graduate decay property of the threshold function, the modifications of wavelet coefficients has shown to affect the smooth area of the denoised image. We cannot discuss image denoising performance without first discussing how to estimate the threshold value.

3.5 Threshold Value

The section title uses threshold value instead of the commonly accepted threshold to avoid readers' confusion about what will be discussed in this section. Section 3.4.1 has discussed various wavelet coefficients threshold functions. However, the application of these threshold functions requires a threshold value. It is the purpose of this section to discuss how to estimate the threshold values, and their application to the threshold functions.

3.5.1 Universal Threshold (Donoho Threshold)

The universal threshold is selected to be discussed in this section because of its influence in the development of wavelet thresholding denoising techniques. The universal threshold estimates one threshold value for the whole wavelet thresholding denoise process. The advantage of this threshold value estimation method is its simple implementation. The universal threshold t_u is first proposed in 1995 by Donoho [17] and is therefore also known as Donoho threshold. It is given by

$$t_u = \sigma_\eta \sqrt{2\ln(M \times N)}, \tag{3.5}$$

where $M \times N$ is the size of the subband image applied to estimate the threshold (i.e. the total number of pixels considered). In practice, σ_η is not given and should be estimated by statistical techniques, such as the method depicted in Equation 3.1; moreover, the original image size is considered such that no information from the required decomposition level is required. The associated MATLAB code is shown in Listing 3.2.1.

Example 3.2 To investigate the effects of threshold function selection in wavelet thresholding denoising, consider the *Sculpture* image corrupted by AWGN with $\sigma_\eta = 25$. The universal threshold is given by

$$t_u = (25) \times \sqrt{2\ln(512 \times 512)} = 124.8832. \tag{3.6}$$

To demonstrate the performance of t_u, we shall consider t that varies over the interval [12.4883, 162.3481] with a step size of $\Delta = 12.4882$ around the universal threshold (i.e. from $0.1t_u$ to $1.3t_u$ with step size of $0.1t_u$). The wavelet image denoising can be easily obtained with a known threshold value by modifying `waveletth` with `thest` replaced by the known threshold value, as shown in the following MATLAB code.

Listing 3.5.1: MSE performance of different wavelet thresholding functions with varying threshold values.

```
>> load('sculpture.mat');
>> [M,N]=size(sculpture);
>> sigma=25; noise = sigma.*(randn([M,N]));
>> f = noise+double(sculpture);
>> J=0; level=3;fsize=M*N;
>> for k=1:13
    thvar=0.1+0.1*(k-1);
    [gh jh thh] = wavelethvar(f, fsize, J, level, thvar);
    gh=imgtrim(gh);
    thhout(k) = thh; mseh(k)=mse(gh,sculpture);
    [gs js ths] = waveletsthvar(f, fsize, J, level, thvar);
    gs=imgtrim(gs);
    thsout(k) = ths; mses(k)=mse(gs,sculpture);
    [gg jg thg] = waveletgthvar(f, fsize, J, level, thvar);
    gg=imgtrim(gg);
    thgout(k) = thg; mseg(k)=mse(gg,sculpture);
>> end
>> plot(thhout,mseh,'-r','LineWidth',1.5);
>> hold on;
>> plot(thsout,mses,'--b','LineWidth',1.5);
>> plot(thgout,mseg,':k','LineWidth',1.5);
>> legend('Hard Threshold', 'Soft Threshold', 'Garrote Threshold')
>> hold off;
```

Three functions wavelethththvar, waveletsthvar and waveleghth-var are modified from waveltth to implement the hard thresholding, soft thresholding, and Garrote thresholding, respectively, where a threshold scaling factor thvar is introduced to vary the threshold value from the range of 0.1× to 1.3× of the computed universal threshold by Equation 3.5. The three MATLAB functions are shown below.

```
function [g j th] = waveleththvar(f, fsize, j, level,thvar)
  th = 0;
  [fll, flh, fhl, fhh] = dwt2(f,'db1','per');
  j=j+1;
  if j<level [fll j th] = waveleththvar(fll, fsize, j, level,thvar);
  else th = thvar*25*sqrt(2*log(fsize));
  end
  flh = hthfun(flh, th); fhl = hthfun(fhl, th); fhh = hthfun(fhh, th);
  g = idwt2(fll, flh, fhl, fhh, 'db1', 'per');
  j=j-1;
end

function [g j th] = waveletsthvar(f, fsize, j, level,thvar)
  th = 0;
  [fll, flh, fhl, fhh] = dwt2(f,'db1','per');
  j=j+1;
  if j<level [fll j th] = waveletsthvar(fll, fsize, j, level,thvar);
  else th = thvar*25*sqrt(2*log(fsize));
  end
  flh = sthfun(flh, th); fhl = sthfun(fhl, th); fhh = sthfun(fhh, th);
  g = idwt2(fll, flh, fhl, fhh, 'db1', 'per');
  j=j-1;
end

function [g j th] = waveletgthvar(f, fsize, j, level,thvar)
  th = 0;
  [fll, flh, fhl, fhh] = dwt2(f,'db1','per');
  j=j+1;
  if j<level [fll j th] = waveletgthvar(fll, fsize, j, level,thvar);
  else th = thvar*25*sqrt(2*log(fsize));
  end
  flh = gthfun(flh, th); fhl = gthfun(fhl, th); fhh = gthfun(fhh, th);
  g = idwt2(fll, flh, fhl, fhh, 'db1', 'per');
  j=j-1;
end
```

Hard, soft, and Garrote threshold functions will be applied to the noisy image and the MSE of the denoised images will be computed. The results are shown in Figure 3.6. These curves clearly indicate that the universal threshold is not “optimal” in the MSE sense. In general, the universal threshold tends to be conservatively high, resulting in over-smoothing the noisy image. This is the case because the derivation of this threshold gives higher priority to ensuring that the denoised image is at least as smooth as the original image

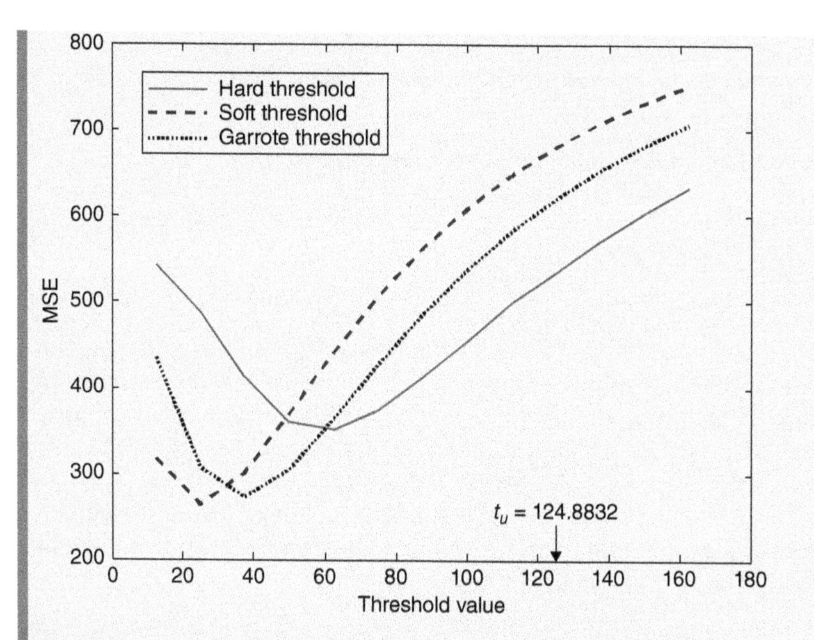

Figure 3.6 The MSE of the wavelet thresholding denoised AWGN corrupted *Sculpture* image with $\sigma_\eta = 25$ with the threshold varies in the interval $[12.4883, 162.3481]$ with a step size of $\Delta = 12.4882$ around the universal threshold $t_u = 124.8832$.

than to minimizing the mean squares error. Often, this threshold is only useful as a starting value when nothing else is known about the image in concern, such as smoothness. One can then estimates a better threshold value depends on the denoising result obtained using the universal threshold. It can be noted that the "optimal" thresholds (the one gives the lowest MSE) for both soft and hard threshold functions are consistently lower than the universal threshold. It is also interesting to note that the "optimal" soft threshold is consistently about half of the "optimal" hard threshold for the noisy *Sculpture* image. At the same time, it can be observed that the optimal Garrote threshold is in between the "optimal" soft and hard thresholds. Actually, this observation is consistent with most natural images and has been widely reported in wavelet thresholding literature, although it has yet to be shown to hold analytically [38].

The function `thest` shown in Listing 3.5.2 is the MATLAB implementation of the universal threshold estimation, where the noise variance σ_η is estimated by Equation 3.1 with the associated MATLAB function `waveletnoiseest` introduced in Listing 3.2.1. Hence, we can compute the required t (i.e. `th`) by applying the function `thest` in the recursive function `waveletth`.

Listing 3.5.2: The threshold value estimation function.

```
function th = thest(fhh,fsize)
  sigma=waveletnoiseest(fhh);
  th = sigma*sqrt(2*log(fsize));
end
```

The variable `fsize` is the size of the data in the wavelet subband to be applied for threshold value estimation. As an example, a 3-level wavelet decomposition of an image of size is 512×1512 will generate a subband image size of 128×128. Noted that Equation 3.5 has to use the image size, and thus `fsize` $= 512 \times 512 = 262144$ is used, instead of the subband image size $128 \times 128 = 16384$. It should be noted that there are more than one method to estimate the value of t. Readers are advised to look up literature to understand other threshold value estimation schemes.

Example 3.3 This example will perform image denoising of all the three wavelet threshold denoising functions, hard thresholding, soft thresholding, and Garrote thresholding, in Section 3.4.1 for the AWGN corrupted *Sculpture* image with $\sigma_\eta = 25$. The MATLAB code in Listing 3.5.3 will call three different wavelet threshold functions, `hthfun`, `sthfun`, and `gthfun` depending on the global variable `whichfun` such that different threshold function will be applied to perform wavelet threshold image denoising. The wavelet decomposition is set to be 3.

Listing 3.5.3: Wavelet threshold image denoising performance with various threshold functions.

```
>> global whichfun
>> [M N]=size(f); fsize = M*N;
>> J=0; level=3;
>> for whichfun=1:3
     [g, j, th] = waveletth(f, fsize, J, level);
     g=imgtrim(g);
     end
```

The threshold function `thfun` in `waveletth` is modified as the following

Listing 3.5.4: Wavelet threshold function selector.

```
function [g] = thfun(f,th)
    % global variable whichfun = 1 -> hard threshold
    % global variable whichfun = 2 -> soft threshold
    % global variable whichfun = 3 -> garrote threshold
    % if not defined -> hard threshold
    global whichfun
    if exist('whichfun','var') wfun=whichfun;
    else wfun=1; end
    if wfun == 2
        g = sthfun(f,th);
    else if wfun == 3
            g = gthfun(f,th);
    else g = hthfun(f,th);
    end
    end
end
```

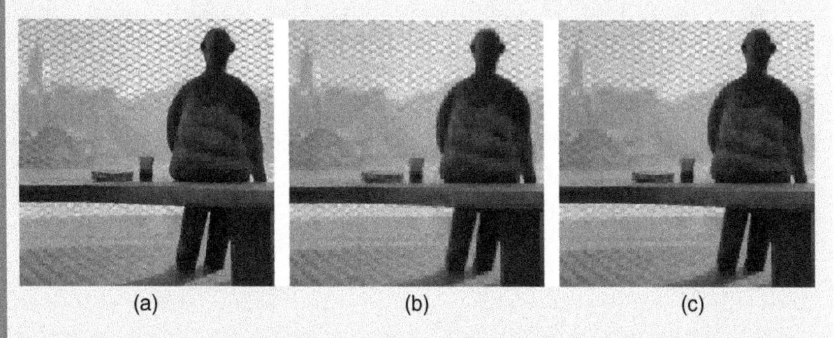

Figure 3.7 Wavelet thresholding denoised *Sculpture* images corrupted by AWGN with σ_η =25, denoised by (a) hard threhsolding (PSNR = 20.7363 dB), (b) soft thresholding (PSNR = 19.7330 dB), and (c) Garrote thresholding (PSNR = 20.1175 dB).

This function will determine which threshold function to call depending on the global variable `whichfun`. The simulation results of the denoised images from an AWGN corrupted image with $\sigma_\eta = 25$ by different threshold functions are shown in Figure 3.7. Further noted that the function `thest` called by each threshold function will generate the universal threshold, which has a value of 129.2228 in this example. Among all three threshold functions, the hard threshold function (result see Figure 3.7(a)) achieves the best PSNR at 20.7363 dB, while the Garrote threshold function achieves the moderate PSNR at 20.1175 dB (result see Figure 3.7(c)), and the soft threshold function achieves the lowest PSNR at 19.7330 dB (result see Figure 3.7(b)). It shows that the threshold function would affect the denoising performance subject to the same threshold value because the image details are mainly composed of high-frequency components and are blended with the noise in the highpass subband image. Different threshold functions cater to the suppression of the small magnitude coefficients in the subband images in different manners, such that different denoise results will be obtained. It can observed that the hard threshold function is more effective in preserving sharp edges (e.g. the outlines of the *Sculpture*) but creating unpleasant discontinuity along soft edges (e.g. meshes) due to the breakpoint problem of the hard threshold. Soft threshold function is able to suppress the discontinuity along soft edges but blurring the outlines of objects (e.g. the *Sculpture*, the book, and the glass) due to constant deviation. The Garrote thresholding gives a balance between two methods.

3.5.1.1 Adaptive Threshold

It has been shown in [17] that under certain restrictive conditions, the universal threshold is optimal. However, in general, the universal threshold is too large which over-smooth the denoised image. This property of the universal threshold is vivid in Example 3.3. This deficiency of the universal threshold can be corrected by making the threshold value to adapt to the level-j of wavelet decomposition with $1 \leq j \leq J$. A particular adaptive threshold value $t_{u,j}$ to apply to level-j threshold function is given by the following function.

$$t_{u,j} = \frac{t_u}{\log(j+1)}, \tag{3.7}$$

such that $t_{u,j}$ is a decreasing function with respect to an increasing j. The associated wavelet threshold function for this adaptive threshold value is implemented as `waveletath` in MATLAB.

Listing 3.5.5: Adaptive threshold.

```
function [g j th] = waveletath(f, fsize, j, level)
  th = 0;
  [fll, flh, fhl, fhh] = dwt2(f,'db1','per');
  j=j+1;
  if j<level [fll j th] = waveletath(fll, fsize, j, level);
  else th = thest(fhh,fsize);
  end
  thj=th/(log(j+1));
  flh=hthfun(flh, thj); fhl=hthfun(fhl, thj); fhh=hthfun(fhh, thj);
  g = idwt2(fll, flh, fhl, fhh, 'db1', 'per');
  j=j-1;
end
```

In a similar manner, another way to adapt the threshold value for the level-j threshold function is to adapt the estimated image noise variance at level-j in the following way.

$$\sigma_{\eta,j} = \frac{\sigma_\eta}{\sqrt{2^{J-j}}}. \tag{3.8}$$

Substitute this into the universal threshold in Equation 3.5, the adapted $t_{u,j}$ will be given by

$$t_{u,j} = \frac{t_u}{\sqrt{2^{J-j}}}. \tag{3.9}$$

This is one of the popular scale-dependent thresholding scheme proposed in [16], known as scale shrink. The associated wavelet threshold function for this adaptive noise variance is implemented as `waveletsth` in MATLAB.

Listing 3.5.6: Scale shrink threshold.

```
function [g j th] = waveletsth(f, fsize, j, level)
  th = 0;
  [fll, flh, fhl, fhh] = dwt2(f,'db1','per');
  j=j+1;
  if j<level [fll j th] = waveletsth(fll, fsize, j, level);
  else th = thest(fhh,fsize);
  end
  thj=th/(sqrt(2^(j-1)));
  flh=hthfun(flh, thj); fhl=hthfun(fhl, thj); fhh=hthfun(fhh, thj);
  g = idwt2(fll, flh, fhl, fhh, 'db1', 'per');
  j=j-1;
end
```

Listed in Table 3.2 are the threshold values computed by Equations 3.5, 3.7, and 3.9, where the σ_η to compute all three threshold values are estimated by

Table 3.2 Wavelet threshold values variations obtained from universal threshold, adaptive threshold `waveletath` and scale shrink `waveletsth` for AWGN corrupted *Sculpture* image with $\sigma_\eta = 25$.

Level J	Threshold value Universal threshold Equation 3.5	Threshold value `waveletath` Equation 3.7	Threshold value `waveletsth` Equation 3.9
3	129.2228	93.2145	64.6114
2	129.2228	117.6236	91.3743
1	129.2228	186.4290	129.2228

Equation 3.8. Please note that the universal threshold estimated by Equation 3.5 with σ_η estimated by `waveletnoiseest` is much bigger than the other two scale-dependent threshold values listed in Table 3.2. It has been shown in Figure 3.6 that the optimal threshold values may be smaller than that of the universal threshold. The following example will investigate the image denoising performance of all three threshold systems listed in Table 3.2.

Example 3.4 The denoising performance of the above-discussed two threshold value adaptation methods can be investigated by executing the function with the noisy image `f` obtained from the *Sculpture* image corrupted by AWGN with $\sigma_\eta = 25$. The hard threshold denoised image with universal threshold is also implemented for comparison.

```
>> load('sculpture.mat');
>> [M,N]=size(sculpture);
>> sigma=25; noise = sigma.*(randn([M,N]));
>> f = noise+double(sculpture);
>> level=3; J=0; fsize=M*N;
>> [g1, j1, th1] = waveletth(f, fsize, J, level);
>> g1=imgtrim(g1);
>> [g2, j2, th2] = waveletath(f, fsize, J, level);
>> g2=imgtrim(g2);
>> [g3, j3, th3] = waveletsth(f, fsize, J, level);
>> g3=imgtrim(g3);
```

The denoised images `g1`, `g2`, and `g3` are plotted in Figure 3.8(a), (b), and (c), respectively, with the corresponding PSNR at 20.7363, 20.9506, and 21.6219 dB, respectively. It can be observed that both scale-dependent

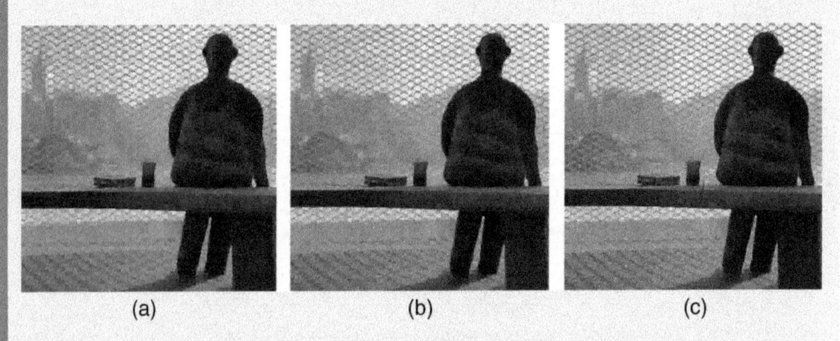

Figure 3.8 Wavelet hard threshold denoising result with adaptive threshold for *Sculpture* image with AWGN at $\sigma_\eta = 25$ with 3-level of decomposition and universal threshold under different threshold value adaptations: (a) universal threshold without adaptation (20.7363 dB); (b) `waveletath` (20.9506 dB), and (c) `waveletsth` (21.6219 dB).

threshold value techniques can produce denoised images with better objective quality than that of the universal threshold value without adaptation. It can be observed that the scale shrink achieves the best performance. Furthermore, it can be observed that Figure 3.8(c) has the least blocking noise, and the background buildings have the cleanest display, where the tallest background building in both Figure 3.8(a) and (b) have been washed out and cannot appreciate the pointy roof-top of the building in both images.

3.6 Wavelet Wiener

The noise residual in Wiener filter denoising result is due to the rigid Wiener template which does not fit for all images. The bigger the template, the smoother the image, but it will also lose more detail texture. While a smaller template will retain the image detail and also the noise. Since the detail coefficients obtained from wavelet transform of the noisy image can be considered as a low-resolution version of the original noisy image; therefore, we can conjecture that applying the Wiener filtering to the low-resolution image with small template size will help to obtain a better denoised image. Furthermore, Section 2.2 has shown that the Wiener filter will remove high-frequency signal components together with the noise. Since the approximate coefficients are mostly high-frequency components, if Wiener filter is being applied, the approximation coefficients will most likely to be removed completely by the Wiener filter. At the same time, since the noise components in the approximate coefficients can be effectively taken care of by thresholding. Such hybrid image denoising method, the wavelet Wiener thresholding method, is implemented in Listing 3.6.1 with scale shrinker. The MATLAB built-in function `wiener` is applied to realize the Wiener filter on detail coefficient images, where the filter size is `[3,3]`.

Listing 3.6.1: Wavelet Wiener threshold hybrid image denoising.

```
function [g, j, th] = waveletwiener(f, fsize, j, level)
  [fll, flh, fhl, fhh] = dwt2(f,'db1','per');
  th=0; j=j+1;
  if j<level [fll, j, th] = waveletwiener(fll, fsize, j, level);
  else th = thest(fhh,fsize);
  end
  thj =th/(sqrt(2^(j-1)));
  flh=hthfun(flh, thj); fhl=hthfun(fhl, thj); fhh=hthfun(fhh, thj);
  if (j==1)
    fll = wiener2(fll,[3 3]);
  end
  g = idwt2(fll, flh, fhl, fhh, 'db1','per');
  j=j-1;
end
```

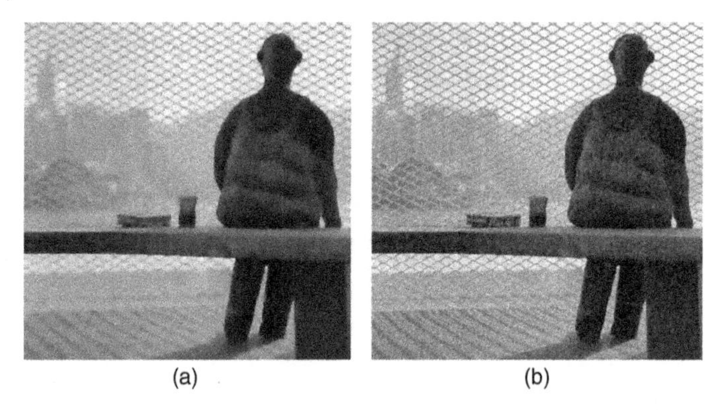

(a) (b)

Figure 3.9 Denoising result on AWGN corrupted *Sculpture* image with σ_η=25: (a) by using Wavelet Wiener hybrid method combining scale shrinker hard thresholding and Wiener filtering (PSNR of 21.5006 dB) and (b) by using Wiener filtering with filter size of 3×3.

The image denoising results obtained by executing `waveletwiener` on AWGN corrupted *Sculpture* image with $\sigma_\eta = 25$ is shown in Figure 3.9(a).

```
>> load('sculpture.mat');
>> [M,N]=size(sculpture);
>> sigma=25; noise = sigma.*(randn([M,N]));
>> f = noise+double(sculpture);
>> level=3; J=0; fsize=M*N;
>> [g, j, th] = waveletwiener(f, fsize, J, level);
>> g=imgtrim(g);
```

The PSNR of the denoised image is 21.5006 dB which is higher than that obtained with all other presented wavelet threshold methods in this chapter. For comparison, the same noisy image is denoised with a 3×3 Wiener filter in spatial domain (method presented in Section 2.2) alone, and the result is shown in Figure 3.9(b). It can be observed that the denoised image obtained by the wavelet Wiener threshold hybrid method has well-defined image edges compared to that obtained by Wiener filter alone (Figure 3.9(b)). Furthermore, the hybrid method (Figure 3.9(a)) effectively alleviated blocking noise in the wavelet hard threshold denoised image. All objects have a smooth texture. Furthermore, the background building is visible in the denoised image. Both the pointy top of the tall building and the arc shape of the convention hall in the background can clearly be observed in this denoised image.

3.7 Cycle Spinning

Except Haar wavelet, almost all wavelet transforms are not translation invariant. It means that the wavelet transform of an image and of a translated version of the image will give different sets of wavelet coefficients. Normally the translation variation properties of the wavelet transform does not affect the denoising performance. However, we can make use of this interesting feature to obtain better denoising result among the local features of the image. It has been discussed in [13] that perfect translation invariance image denoising can be obtained by linear combination of all possible translation wavelet denoising results. In particular, the translations are accomplished by cycle spins, such that the translated image has the same size as that of the original image, and at the same time has the same information content. Consider translating the image to the left hand side by 1 pixel. When the translation operator hits the image boundary, various boundary extension techniques can be applied, and cycle spinning will apply periodic extension such that the 1 pixel that is off the left-hand side boundary will be appended to the right-hand side boundary similar to that of the periodic extension. Each shift generated by cycle spinning will be denoised and then shifted back to the original image positions and then averaged to generate the final denoised image. For an $M \times N$ image, there are $M \cdot N$ different cycle spins. It is obvious that using all cycle spins is impossible in practice. Fortunately, the number of translations required to provide a good approximate invariance denoising result is usually much lower than $M \cdot N$.

Despite all the nice features of the cycle spinning image translation, it has one major disadvantage which sparks the research community to look for other translation image generation methods. This disadvantage is the discontinuity induced by the cycle spin due to periodic extension alike translation, which will severely affect the image denoising performance. In this book, we use a different extension scheme to alleviate the boundary discontinuity problem. After shifting the image to the left-hand side, the missing pixel at the right-hand side image boundary will be filled up with the pixels obtained by symmetric extension. Similarly, the same technique is applied when the image is shifted up. The following MATLAB script will generate the translated image `fk` from the image `f` with size $M \times N$ by shifting the image `f` with an offset of `[-k, -k]` using MATLAB function `circshift`, as shown in Figure 3.10(a). Part of `fk` is formed by the shifted version of `f`, where the remaining pixels are obtained by flipping a copy of selected pixels from `fk` to maintain the continuity. The MATLAB code for the boundary extension is illustrated in Figure 3.10. The MATLAB code of the shifting is shown below.

```
fk = circshift(f,[-1*k, -1*k]);
fk(M-k+1:end,:) = flipud(fk(M-2*k+1:M-k,:));
fk(:,N-k+1:end) = fliplr(fk(:,N-2*k+1:N-k));
```

Wavelet threshold denoising will be applied to each translated image with different translations. The denoised image will be translated back to the original position. However, unlike cycle spinning, there will be missing pixels around the shifted image (k pixels in the both left and right, and also k pixel in both top and

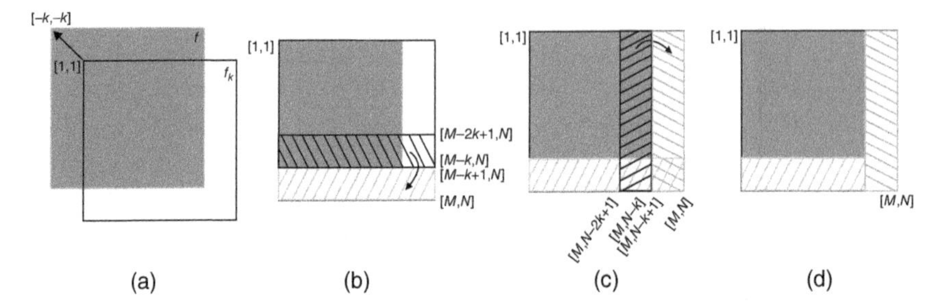

Figure 3.10 Boundary extension in the function `cyclespin`: (a) The formation of translated image `fk` from the image `f` by the MATLAB function `circshift`, where the image `f` is shifted with offset of `[-k,-k]`. (b) The pixels `fk(M-2*k+1:M-k,:)` is flipped and copied to fill up the bottom part of `fk` to maintain the image continuity. (c) The pixels `fk(:,N-2*k+1:N-k)` is flipped and copied to fill up the right part of `fk` to maintain image continuity. (d) The final `fk` is with the same size of [M,N].

bottom). This void will be filled by pixels obtained from another denoised image, such as the one obtained from `f` directly without translation. The following MATLAB code will translate the denoised image `gk` back to the centre and fill the void with pixels from another denoised image `g0` to obtain the final denoised image `gt`.

```
gt = g0;
gt(k+1:M-k,k+1:N-k)=gk(1:M-2*k,1:N-2*k);
```

The function `cyclespin` listed in Listing 3.7.1 puts the above together to form an approximate translation invariance wavelet scale shrink threshold denoise function with k translations.

Listing 3.7.1: Cycle spinning.

```
function g = cyclespin(f,level,spinsize)
[M,N]=size(f);
fsize=M*N;
[g0,j0,th0]=waveletsth(f,fsize,j,level);
g=g0;
for k=1:spinsize
    fk = circshift(f,[-1*k, -1*k]);
    fk(M-k+1:end,:) = flipud(fk(M-2*k+1:M-k,:));
    fk(:,N-k+1:end) = fliplr(fk(:,N-2*k+1:N-k));
    [gk,jk,thk] = waveletsth(fk,fsize,0,level);
    gt = g0;
    gt(k+1:M-k,k+1:N-k)=gk(1:M-2*k,1:N-2*k);
    g=g+gt;
end
g=g/(spinsize+1);
end
```

The advantage of this method is that it alleviates the noise induced by the boundary discontinuity problem after cycle spinning, which is traded with an increase in residual noise when packing the voids in the denoised image after translating back to the original position. As a result, a breakeven on the number of translations is expected in `cyclespin`. The following example will investigate the influence of the number of translation k on the wavelet hard threshold image denoising objective performance using PSNR.

Example 3.5 This example will apply `cyclespin` with `spinsize` translations to obtain a set of denoised images, where `spinsize` varies from 1 to 40. The noisy image considered in this example is the AWGN corrupted *Sculpture* image with $\sigma_\eta = 25$. The PSNR of denoised image is recorded in the vector `ps`. Finally, `ps` is plotted in Figure 3.11. It can be observed that the PSNR increases with increasing number of translations, also known as spins, until it reaches a maximum at `spinsize=7`. After that the PSNR drops linearly with an increase number of spins. This observation is consistent with the fact that the trade-off between the increase in residue noise induced to the denoised image by filling the voids of the denoised image translated back to the original position has reached a breakeven point at `spinsize=7` with the noise reduction gain obtained by averaging different spinned denoised images. The image considered in this example has a size of 512 × 512. When we further increase the number of spin to 8, i.e. `spinsize=8`, the number of pixels in the void

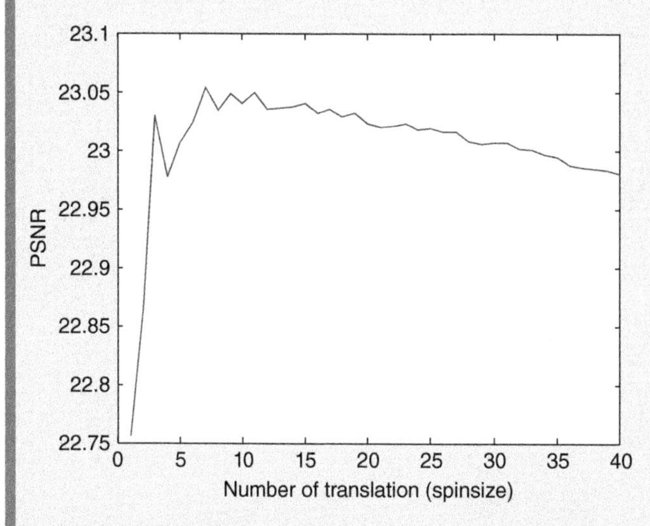

Figure 3.11 The influence of the number of translations (`spinsize`) on the wavelet scale shrink threshold denoising on the image quality.

will be $2 \times 8 \times 512 + 2 \times 8 \times 512 - (2 \times 8)^2 = 16128$, which is 6.15% of the total number of pixels. It is thus conjectured that the residue number in this 6% of total number of pixels that was used to fill the void, which is significant enough to reduce the PSNR of the final denoised image.

Listing 3.7.2: Cycle spinning image denoising example with varying spin size.

```
load('sculpture.mat');
[M,N]=size(sculpture);
sigma=25; noise = sigma.*(randn([M,N]));
f = noise+double(sculpture);
level=3; J=0; fsize=M*N;
ps=zeros(40,0);
for spinsize = 1:40
g=f;
g = cyclespin(f,level,spinsize);
g=imgtrim(g);
ps(spinsize) = psnr(g,double(sculpture));
end
plot(1:spinsize,ps);
```

Knowing that `spinsize=7` will produce the best denoised image. The following example will apply `cyclspin` to image denoising.

Example 3.6 Consider the wavelet hard thresholding image denoising with seven spins for the AWGN corrupted *Sculpture* image with $\sigma_\eta = 25$ using `cyclespin`. The following MATLAB code is applied to obtain the denoised image in Figure 3.12.

Figure 3.12 Wavelet hard threshold denoising with scale shrink threshold and seven cycle spinnings where the PSNR of the denoised image is 23.0543 dB.

Listing 3.7.3: Cycle spinning image denoising example.

```
load('sculpture.mat');
[M,N]=size(sculpture);
sigma=25; noise = sigma.*(randn([M,N]));
f = noise+double(sculpture);
level=3; J=0; fsize=M*N;
spinsize=7;
g = cyclespin(f,level,spinsize);
g=imgtrim(g);
```

The objective performance of the denoising method in Listing 3.7.3 is very good, where the PSNR is 23.0543 dB which is the highest for all the wavelet threshold denoising methods discussed in this chapter. The denoised image is shown in Figure 3.12. It is vivid that the subjective performance of the denoised image is also very good. The objects in the image have sharp edges, and the texture areas are smooth. The metal mesh is sharp and there is only minimal residual noise detected around the mesh. The appearance of the background buildings is almost as clear as the noise-free image. We can therefore make two conjectures from this example. First, averaging denoised image with different additive noise properties help to produce very good denoised final image. This property will be further investigated in Chapter 7. Second, the wavelet threshold artifacts are translation dependent. The artifacts are connected in some way with the precise alignments between the signal features with the basis function features (i.e. the subband filters). A correct translation will help to align the signal features and basis function features to generate fewer of the artifacts. However, there is no guarantee on which translation will be the best translation. As a result, averaging the image denoising results from different translation will help to average out the artifacts obtained in different denoised image, which will reduce the signal power of the artifacts when the artifacts do not align.

3.8 Fusion

It has been shown in Section 2.3.1 that summing up noisy images obtained from the same image corrupted by AWGN generated with identical but independent processes can denoise an image. The same technique has been applied in Section 3.7 and has shown that better quality image denoising results can be obtained. The next question regarding the application of this method will be how to combine the images together to obtain a better denoised image besides averaging. The DWT provided the right mean to perform a different way to

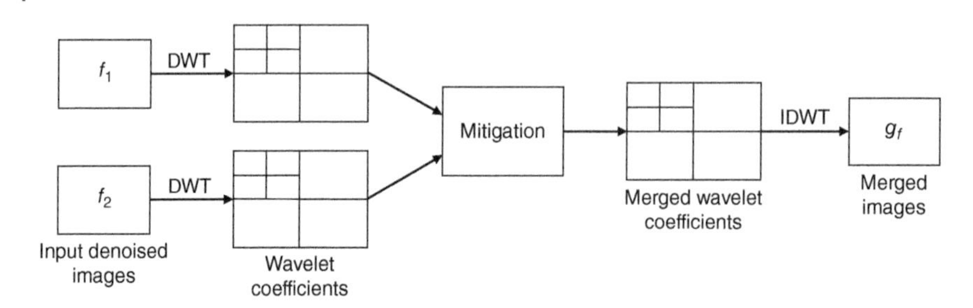

Figure 3.13 DWT fusion: wavelet based image combining algorithm.

combine these images to obtain a better denoised image. To understand how the DWT does that, let us consider two denoised images f_1 and f_2. The way that these two images combine together by DWT is known as DWT fusion which is a process to combine information from multiple images of the same scene. The result of image fusion is a new image that retains the most needed information and characteristics of each input image. The flow diagram of DWT fusion is shown in Figure 3.13. The details of wavelet image fusion starts with two source images f_1 and f_2, where DWT will be applied to both images to obtain the approximation and detailed subband coefficients at the required level. The approximation and detailed subband coefficients of both images are combined by using merging technique. The merged wavelet coefficients will be applied with inverse DWT to generate the merged image g in spatial domain.

Image fusion is the mixing of two or more images together to create a new image. Denoising makes use of image fusion to fuse two or more image denoising results obtained via different methods to create a single denoised image which hopefully will better resemble the noise-free image than any of the denoised images at the input of the fusion algorithm. Image fusion in wavelet domain will perform separate operations in the detailed and approximation coefficients of the wavelet transformed images. The possible operations include mean, max, and min mathematical operations. As a result, there are a total of nine combinations of operations on the approximation and detailed wavelet coefficients. Among all nine different operations, the fusion based image denoising method will apply the mean operation to the detail image and the max operation to the approximation coefficients which has been demonstrated to be able to achieve the best performance. The reason is simple, the detail coefficients are similar to a downsampled version of the denoised images at the input of the algorithm. It has been shown in Section 2.1 that the mean operation can smooth image noise. At the same time, we have shown in Section 3.4 that approximation coefficients with small magnitudes are highly probable to be contributed by the noise. Therefore, in order to avoid the noise effect in the fused image, an obvious choice is to choose the largest approximation coefficients among approximation coefficients obtained from all the input denoised images.

Image fusion is preferred in high-frequency domain. This is based upon the perception that human visual system is more sensitive to local luminance contrast. Number of decomposition levels required depends upon the resolution demanded. The fusion procedure contains the following steps

3.8.1 Baseband Image Fusion

The detail wavelet coefficients are also known as the baseband image. The baseband image are lowpass information of the input images. Most of the energy in an image is concentrated in the low-frequency region. Typically, an image is supposed to have a spectrum that decays with increasing frequency (as shown in Figure 2.2). Different baseband fusion techniques are available in literature, which includes some of the techniques discussed in the following.

3.8.1.1 Simple Average

The baseband image of the fused image is obtained from the averages of the baseband images of the two input images. Mathematically,

$$g_{LL}^{j} = \frac{f_{1,LL}^{j} + f_{2,LL}^{j}}{2}, \tag{3.10}$$

where f_1 and f_2 are the two input images, and g is the fused image. The above average operations will be performed across all $1 \leq j \leq J$ levels of the wavelet decomposed subband image sets. This approach provides poor contrast in the fused image. However, if the input images have similar contrast, the image contrast will be maintained in the fused image. Therefore, it has been very popular in image fusion application because of its simplicity.

3.8.1.2 Arithmetic Combination

The arithmetic combination method was first proposed in [41], where the fused baseband image is given by

$$g_{LL}^{j} = f_{1,LL}^{j} + f_{2,LL}^{j} - \frac{\mu_{f_{1,LL}^{j}} + \mu_{f_{2,LL}^{j}}}{2}, \tag{3.11}$$

where $\mu_{f_{1,LL}^{j}}$ and $\mu_{f_{2,LL}^{j}}$ are the mean values of the two baseband images of the input images f_1 and f_2, respectively. Like other arithmetic fusion, the fusion defined above is susceptible to destructive superposition, especially when the baseband images have opposing illumination levels. The fused images are usually observed to look very natural because the amount of zero mean information is low.

3.8.1.3 Correlation Base

This method was first proposed in [6], where correlation is applied as the selection criteria. If the correlation among images is more than a certain threshold, then images are averaged out. Otherwise, a weighted combination is used based on energy proposition. This method is more effective for images with similar contents, such as denoised images of the same scene. In fact, for single image denoising problem, we expect almost all denoised images will fall into the "average" criteria, and therefore, the correlation base fusion method will produce the same g_{LL} as that obtained by the baseband average image based fusion method.

3.8.2 Detail Images Fusion

The detail subband images consist of the highpass information of the image. Performance of fusion methods mainly depends on how well the highpass information is transferred to the fused image. This is because the human visual system is more sensitive to edge information. Various approaches for detail images fusion are discussed in the following.

3.8.2.1 Simple Average

Similar to the simple average in baseband image fusion, the simple average for all three details coefficients images are shown in the following

$$g^j_{LH} = \frac{f^j_{1,LH} + f^j_{2,LH}}{2}, \tag{3.12}$$

$$g^j_{HL} = \frac{f^j_{1,HL} + f^j_{2,HL}}{2}, \tag{3.13}$$

$$g^j_{HH} = \frac{f^j_{1,HH} + f^j_{2,HH}}{2}. \tag{3.14}$$

This method provides poor visual quality of the fused image with blurred edges.

3.8.2.2 Select Max

This method was first proposed in [35], which considers the absolute value of the coefficient as an indication of saliency. Preference is given to a pixel with more saliency, and hence, the fusion is performed with the following functions

$$g^j_{LH}[m,n] = \begin{cases} f^j_{1,LH}[m,n], & \text{if } |f^j_{1,LH}[m,n]| > |f^j_{2,LH}[m,n]|, \\ f^j_{2,LH}[m,n], & \text{Otherwise.} \end{cases} \tag{3.15}$$

$$g^j_{HL}[m,n] = \begin{cases} f^j_{1,HL}[m,n], & \text{if } |f^j_{1,HL}[m,n]| > |f^j_{2,HL}[m,n]|, \\ f^j_{2,HL}[m,n], & \text{Otherwise.} \end{cases} \tag{3.16}$$

$$g^j_{HH}[m,n] = \begin{cases} f^j_{1,HH}[m,n], & \text{if } |f^j_{1,HH}[m,n]| > |f^j_{2,HH}[m,n]|, \\ f^j_{2,HH}[m,n], & \text{Otherwise.} \end{cases} \tag{3.17}$$

The above maximum selection operation is performed across all wavelet coefficients in all three details coefficients images. The fused image obtained from this approach usually has sharp edges and better visual quality in the texture region.

3.8.2.3 Cross Band Fusion

To further improve the visual quality of the edges of the fused image, the cross band information should be considered at the same time. the cross band fusion proposed in [41] compares and selects the one with larger sum of magnitude of all three detail subband images between the two source images to present in the fused image. The detail mathematical formulation of this image fusing method is given by

$$\begin{array}{lcl} \text{If} \begin{array}{c} |f^j_{1,LH}[m,n]| + |f^j_{1,HL}[m,n]| + |f^j_{1,HH}[m,n]| \\ > \\ |f^j_{2,LH}[m,n]| + |f^j_{2,HL}[m,n]| + |f^j_{2,HH}[m,n]| \end{array} & \rightarrow & \begin{cases} g^j_{LH} = f^j_{1,LH}, \\ g^j_{HL} = f^j_{1,HL}, \\ g^j_{HH} = f^j_{1,HH}. \end{cases} \\ \text{Otherwise} & \rightarrow & \begin{cases} g^j_{LH} = f^j_{2,LH}, \\ g^j_{HL} = f^j_{2,HL}, \\ g^j_{HH} = f^j_{2,HH}. \end{cases} \end{array} \tag{3.18}$$

The fused image obtained by this method can sometime achieve image with edges as sharp as that obtained from the select max approach. The added benefits of this approach is that the ringing noise is alleviated in the fused image obtained by this method when compared to that obtained by the select max approach.

Example 3.7 This example will develop an image fusion function.

Listing 3.8.1: Wavelet image fusion example.

```
>> [M,N]=size(f1);
>> fsize=M*N;
>> [g,j,th] = waveletfus(f1,f2,fsize,0,3);
```

The `waveletfus` will fuse two images `f1` and `f2` with the same size $M \times N$ into a single image `g` using baseband average and select max detail images. Wavelet based denoising using Garrote threshold with adaptive universal threshold value is applied to denoise `f1` and `f2` before fusing them together. The Garrote threshold function is applied to all the detail images before fusion by select max approach. The threshold function will help to remove subband image noise before fusion. This will help to remove outliers before the Select Max approach and hence produce a better fused images. The MATLAB code of `waveletfus` function is listed in Listing 3.8.2.

Listing 3.8.2: Two images fusion.

```
function [g,j,th] = waveletfus(f1,f2,fsize,j,level)
  [f1ll, f1lh, f1hl, f1hh] = dwt2(f1,'db1','per');
  [f2ll, f2lh, f2hl, f2hh] = dwt2(f2,'db1','per');
  th=0; j=j+1;
  if j<level
    [fll,j,th]=waveletfus(f1ll,f2ll,fsize,j,level);
  else
    th1 = thest(f1hh,fsize);
    th2 = thest(f2hh,fsize);
    th = min(th1,th2);
  end
  thj =th/(log(j+1));
  f1lh = gthfun(f1lh,thj); f1hl = gthfun(f1hl,thj);
  f1hh = gthfun(f1hh,thj); f2lh = gthfun(f2lh,thj);
  f2hl = gthfun(f2hl,thj); f2hh = gthfun(f2hh,thj);
  flh = maxmag(f1lh,f2lh); fhl = maxmag(f1hl,f2hl);
  fhh = maxmag(f1hh,f2hh); fll = (f1ll+f2ll)/2;
  g = idwt2(fll,flh,fhl,fhh,'db1','per');
end
```

The select max operation on the detail coefficients images are performed by the function `maxmag`. The function `maxmag(a,b)` returns a matrix with its entry contains the coefficients from either `a` or `b` at the same entry location that has the largest magnitude. The implementation of `maxmag` is shown below.

Listing 3.8.3: Maximum magnitude selection function.

```
function c = maxmag(a,b)
[m,n]=size(a);
c=zeros([m,n]);
for i=1:m
    for j=1:n
        d = abs(a(i,j));
        e = abs(b(i,j));
        if d>e c(i,j) = sign(d)*d;
        else c(i,j) = sign(e)*e;
      end
    end
end
```

Let's consider two noisy images obtained by corrupting *Sculpture* image with AWGN with $\sigma_\eta = 25$, which can be generated by the following MATLAB script.

```
>> load('scultpure.mat');
>> [M,N] = size(scultpure);
>> sigma=25;
>> f1 = sigma.*(randn([M,N]))+double(sculpture);
>> f2 = sigma.*(randn([M,N]))+double(sculpture);
```

The PSNR of these two noisy images are 20.1793 and 20.1634 dB. This two images can be fused together by `waveletfus` using the following MATLAB script

```
>> [M,N]=size(f1);
>> fsize=M*N;
>> [g,j,th] = waveletfus(f1,f2,fsize,0,3);
>> g=imgtrim(g);
```

(a)

(b)

Figure 3.14 Wavelet fusion: (a) one of the input image (f_1) applied in wavelet fusion (a) AWGN corrupted *Sculpture* image with $\sigma_\eta = 25$ and PSNR = 20.1793 dB, and (b) the denoised result of the wavelet fusion where the PSNR = 21.9993 dB.

The PSNR of the fused image `g` is 21.9993 dB, which shows that fusing the two AWGN corrupted images together will help to denoise the image, an observation that is also detected in Section 3.7. The fused image is shown in Figure 3.14(b). One of the noisy image `f1` is shown in Figure 3.14(a) for comparison. Both the subjective quality and the objective quality of the fused image are better than that of either noisy images. In particular, the residue noise in the fused image is very low, and the objects in Figure 3.14(b) have sharp edges, and smooth texture areas. The background buildings are very clear. The reader might observe ghost edges in the fused denoised image. This is because of the edges of individual denoised image have different dislocation caused by the noise. When fused together, because the edge signals are strong signals, therefore, different dislocated edges will survive and pass through the fusion operations. As a result, there are cases where more than one line will be observed along the object edges, which is known as the ghost edges.

3.9 Which Wavelets to Use

Only the Haar wavelet `'db1'` is considered in this chapter. Readers may ask if other wavelets give better image denoising result? To answer this question, we shall have to understand that in wavelet signal analysis, the signal is represented in terms of its coarse and detail approximation at scale J [48]. The key to efficient multiresolution signal representation by wavelet depends on the property of the wavelet basis. The three key properties of the wavelet bases are

Regularity: The regularity of the scaling function (the converged function obtained by convoluting the lowpass filter by itself infinite times) has mostly a cosmetic influence on the error introduced by thresholding or quantizing the wavelet coefficients. If the scaling function is smooth, then the generated error is a smooth error. For image denoising applications, a smooth error is often less visible than an irregular error. Better quality images are obtained with wavelets that are continuously differentiable than those obtained from the discontinuous Haar wavelet. Wavelet regularity increases with the number of vanishing moments. As a result, choosing high regularity wavelet is the same as choosing wavelets with large vanishing moments.

Number of vanishing moments: This affects the amplitude of the wavelet coefficients at fine scale. For smooth regions, wavelet coefficients are small at fine scales if the wavelet has enough vanishing moments to take advantage of the image regularity. A wavelet has m vanishing moments if and only if its scaling function can generate polynomial of degree smaller than or equal to m. Both the number of vanishing moments and the regularity of orthogonal wavelets are related, but it is the number of vanishing moments and not the regularity that affects the amplitude of the wavelet coefficients at fine scales [38].

Kernel sizes: These need to be reduced to minimize the number of high amplitude coefficients. On the other hand, a large kernel size is required to provide enough vanishing moments. Therefore, the choice of optimal wavelet is a trade-off between the number of vanishing moments and kernel size.

Every function that satisfies the admissibility condition can be used in a wavelet transform and generates its own wavelets. The ability to approximate signals with a small number of non-zero coefficients is undoubtedly the key to the success of wavelets for image denoising. An orthonormal function should also be used such that individual subband signals are independent of each others, and hence, the denoising errors will not propagate across subbands. As a result, orthonormal wavelet with the wavelet transform of an image produces few coefficients, which are independent of each subbands, are the best wavelet for image denoising. On the other hand, image denoising should preserve phase information, as human visual system is sensitive to phase error. Since orthogonal wavelets do not have linear phase, except Haar wavelet. Therefore, biorthogonal wavelets with linear phase transform kernel will be the best engineering choice for image denoising.

In summary, the following listed the important properties of wavelet functions in image denoising

1) Compact support (as compact kernel size, which minimizes the high-frequency artifacts in the interpolated image, and also provides efficient implementation)
2) Symmetry (useful in avoiding phase noise in image processing)
3) Orthogonality (reduce noise propagation across subbands)
4) Regularity and degree of smoothness (improve the smoothness of the denoising error and hence less visible)

One should choose the wavelet filters based upon the characteristics of the image for suitable perceptual quality. However, the underlying operations in the algorithm are not image content dependent and hence non-adaptive in nature. As a result, should we spend time to choose a specific wavelet with the goal of finding the optimal wavelet for a given image, and risking of losing generality? The readers may want to adapt the wavelet image denoising algorithms discussed in this chapter for image denoising. But the authors' trial showed that the use of "turnkey" resources provided by MATLAB, in particular the Haar wavelet, will provide just as good image denoising results as other wavelet transformations.

3.10 Summary

In this chapter, we have reviewed the efficiency of wavelet transform in the representation of an image. Through wavelet decomposition, signals in different natures representing different image features, can be approximated separately into coarse or detail coefficients without the hampers of feature localization problems as observed in other transform domain interpretation. Similar to other transform domain denoising techniques, noise suppression through thresholding is still the key of the wavelet transform based image denoising. However, due to the high regularity and integrity of the approximation, the wavelet transform based approaches outperform the others in alternative domains, in preserving the image features and also alleviating the blocking artifacts. Moreover, the decomposition property of the wavelet transform facilitates the application of adaptive threshold values to enhance the noise removal performance, where the threshold values can be adjusted according to the level of the decomposition. The other important property of the wavelet transform, the translational property, allows advance noise removal by cyclic spin adaptive wavelet thresholding method. It is vivid that this method is able to retain image details, which is useful in preserving the high-frequency components in the image. Finally, the wavelet transform is an effective tool for averaging out noise through wavelet fusion. The averaging is realized in the decomposition levels such that only the noise concentrated subband images would be altered. Rectified subband images are fused and reconstructed to form the final denoised image. The fusion concept lays a useful framework for other advance image denoising techniques, which will be discussed in Chapters 4 and 7. Wavelet transform is a comprehensive topics. We encourage readers to go deep into it and to explore other possibilities in applying wavelet transform in image denoising. Though wavelet transform based method shows pleasant denoising results, it is not suitable for all sort of noises, e.g. salt and pepper noise, which leads us the investigation on projection based and optimization based denoising method to be discussed in Chapters 4, 5, and 6.

Exercises

3.1 Study and comment the influence of the number of translations in cycle spinning with wavelet denoising using Haar wavelet and soft thresholding.

3.2 Prove that if the variance of any windowed section of a signal is a constant, then the variance of the wavelet transform is a constant in each resolution level. Is this true for any wavelet transform?

3.3 This exercise will investigate the effect of the Wavelet basis in denoising performance.

1. Create a AWGN corrupted *Sculpture* image with $\sigma_\eta = 10$. Display the 400th column of both the noise-free and the noisy *Sculpture* images.
2. Use the following MATLAB function with `th=100` and `level=3` to perform wavelet hard thresholding on the noisy image in exercise 3.3.1.

```
function g = waveleththth(f, th, level)
  [c,s] = wavedec2(f,3,'db1');
  c = hthfun(c,th);
  g = waverec2(c,s,'db1');
end
```

Please record the PSNR of the denoised *Sculpture* image, and plot column 400.

3. Modify the above MATLAB function to use `'db2'` and perform the denoising with the PSNR of the denoised image recorded together with the line graph of the 400th column of the denoised image plotted in a new figure.
4. Modify the above MATLAB function to use `'db4'` and perform the denoising with the PSNR of the denoised image recorded together with the line graph of the 400th column of the denoised image plotted in a new figure.
5. Modify the above MATLAB function to use `'db8'` and perform the denoising with the PSNR of the denoised image recorded together with the line graph of the 400th column of the denoised image plotted in a new figure.

Compare and discuss the 400th column of the denoised images obtained by `'db1'`, `'db2'`, `'db4'`, and `'db8'`. What's the impact to the denoised image with different wavelet functions. Finally, what is the physics on the performance obtained by each wavelet functions, and among this four wavelet functions, which one will you choose for image denoising.

4

Rank Minimization

The sparse representation of an image through DCT and its application in denoising has been demonstrated in Section 2.3. We have also discussed wavelet transform as a better alternative to DCT in image denoising. In particular, the wavelet transform helps to provide certain adaptation to the complex image structure. The key to construct effective denoising method is on how to make full use of the prior information in the image. This prior information leads to the rank minimization method which has been applied in a wide range of applications such as system identification, computer vision, and machine learning [21]. In this chapter, we shall discuss the development of the matrix rank minimization-based image denoising methods.

The rank minimization problem is generally formulated as

$$\min_{g} \quad \frac{1}{2} \|g - f\|_F^2 + \lambda \text{rank}(g), \tag{4.1}$$

where rank(⋅) is the rank of the image $g \in \mathbb{R}^{M \times N}$ which is the unknown low-rank matrix. In general, the rank minimization problem is both NP-hard and non-convex. Fortunately, the problem in Equation 4.1 can be solved with a hard thresholding operator on the singular values of the noisy image f, such that

$$g = U\mathcal{T}(\Sigma, t)V^T, \tag{4.2}$$

where $f = U\Sigma V^T$ is the *singular value decomposition* (SVD) of f, and $\mathcal{T}(\cdot, t)$ is the threshold function defined in similar manner as that presented in Chapter 3 with threshold value t.

Digital Image Denoising in MATLAB, First Edition. Chi-Wah Kok and Wing-Shan Tam.

Companion website: www.wiley.com/go/kokDeNoise

4.1 Singular Value Decomposition (SVD)

It has been demonstrated in Section 2.1 that natural image is bandlimited, and thus after Fourier transformation, only a few low-frequency basis functions will suffice to completely describe the image. In order to find the best orthogonal transformation that can describe the image with as few basis functions as possible, we can use the mathematical tool known as *singular value decomposition* (SVD) which will decompose a rectangular matrix $f \in \mathbb{R}^{M\times N}$ into products of three matrices, an orthogonal matrix U, a diagonal matrix Σ, and the transpose of another orthogonal matrix V.

$$f = U \times \Sigma \times V^T, \tag{4.3}$$

where $U^TU = I_M$, $V^TV = I_N$, and Σ is a diagonal matrix of size $\min(M, N)$. MATLAB has an efficient built-in function `svd` to compute the SVD as shown in Listing 4.1.1.

Listing 4.1.1: SVD an image.

```
>> [U S V] = svd(f,0);
```

Without loss of generality, and to simplify our discussions, we shall consider the case of the noisy image $f \in \mathbb{R}^{M\times M}$, such that $\Sigma \in \mathbb{R}^{M\times M}$ is a diagonal matrix of the singular values s_ℓ, $\ell = 1, \dots, M$. Furthermore, the singular value s of f is assumed to satisfy

$$s_1 \geqslant s_2 \geqslant \cdot \geqslant s_k \geqslant s_{k+1} = \cdots = s_M = 0, \tag{4.4}$$

such that f is also known as a rank k matrix. Therefore, some of the image denoising method will rewrite the rank(f) in Equation 4.1 as

$$\|f\|_* = \sum_{\ell=1}^{k} s_\ell, \tag{4.5}$$

where $\|f\|_*$ is known as the nuclear norm. The replacement of rank(f) with the nuclear norm makes the optimization problem in Equation 4.1 more tractable, but the problem is still NP-hard and non-convex.

To understand how to apply SVD to perform denoising, let's consider a plot of the squares of the singular values of the noise-corrupted image f. The squares of the singular values of the noise η applied on f will be plotted in the same figure. Similarly, the squares of the singular values of the *Sculpture* image are also plotted in the same figure. This plotting can be obtained by the following MATLAB function.

Listing 4.1.2: Plot of the squares of singular values.

```
>> sigma=50;
>> noise = sigma.*(randn([size(sculpture)]));
>> f=noise+double(sculpture);
>> [u s v] = svd(f,0);
>> s2 = diag(s).^2;
>> figure(1); semilogy(s2, 'r', 'LineWidth',1.5);
>> [un sn vn] = svd(noise,0);
>> sn2 = diag(sn).^2;
>> figure(2);
>> semilogy(s2,'r','LineWidth',1.5);
>> hold on;
>> semilogy(sn2,'--b','LineWidth',1.5);
>> hold off;
>> [us ss vs] = svd(double(sculpture),0);
>> ss2 = diag(ss).^2;
>> figure(3); semilogy(ss2,':g', 'LineWidth',1.5);
>> hold on;
>> semilogy(s2,'r','LineWidth',1.5);
>> semilogy(sn2,'--b','LineWidth',1.5);
>> hold off;
```

The above MATLAB source plotted three figures. The first figure, Figure 4.1(a) plotted the squares of all 512 singular values of the AWGN corrupted *Sculpture* image with $\sigma_\eta = 50$. Noted that even though the y-axis of the figure is in log scale, it is still vivid that the curve starts with a sharp decay. It then turns to a very slow decay for almost three quarters of the whole curve. Finally it drops sharply by the end of the curve. To understand why Figure 4.1(a) behaves like that, we have to consider Figure 4.1(b) which plotted both the squares of the singular values of the AWGN corrupted *Sculpture* image (solid line) and the squares of the

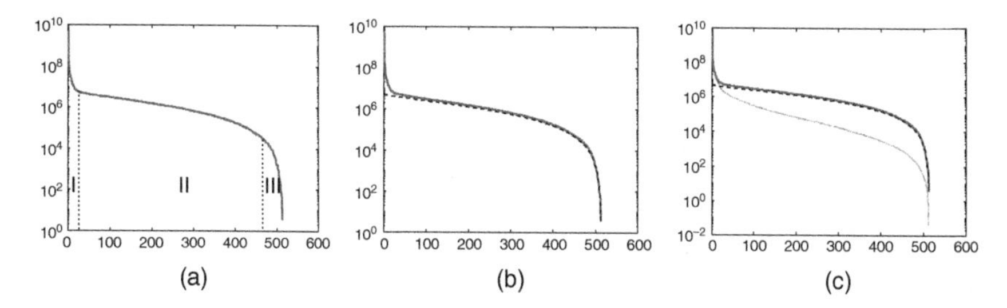

Figure 4.1 Squares of the singular values of (a) AWGN corrupted *Sculpture* image with $\sigma_\eta = 50$; (b) AWGN corrupted *Sculpture* image with $\sigma_\eta = 50$ and the AWGN; and (c) AWGN corrupted *Sculpture* image with $\sigma_\eta = 50$ and the AWGN and the noise-free *Sculpture* image.

singular values of the AWGN (dashed line). It can be observed that the Part II and the Part III parts of the curve in Figure 4.1(a) follow closely with that of the noise signal. In other words, it can be conjecture that these singular values of the AWGN corrupted *Sculpture* image are almost surely contributed by the noise. In theory, projecting AWGN onto any kind of orthonormal system will obtain a set of projected coefficients, where all the coefficients will have the same power. This is the reason why the Fourier spectrum of AWGN signal is a flat line. However, the squares of the singular values are not observed to have the same magnitude in Figure 4.1(b) but a monotonic decreasing property. This is because of the order statistics property of the singular values of the Gaussian random matrix that even random numbers will be observed to be monotonic after they have been sorted. The squares of the singular values of the noise-free *Sculpture* image in Figure 4.1(c) (dotted line) help to confirm that part I of the curve in Figure 4.1(a) for the squares of the singular values of the AWGN corrupted *Sculpture* image is mostly contributed by the image signal. Even though we expect natural images to have low-rank, and thus there should be limited number of non-zero singular values. However, it can be observed that a large part of the noise-free image curve in Figure 4.1(c) has values lower than 10^6 which is an order of magnitude smaller than the squares of the singular values in the AWGN curve (dashed line) but still not zero. Actually, these singular values in the noise-free image curve (dotted line) are the results of the background noise in the *Sculpture* image, which exists in all natural images, no matter how small it is. Since it is an order of magnitude smaller than that of the AWGN (dashed line), it will not affect the quality of the denoised image even if they are ignored in the denoising algorithm.

4.2 Threshold Denoising Through AWGN Analysis

An intuitive observation from Figure 4.1(c) is that the noise signal and the corresponding noisy image have interception. On the left-hand side of the interception, the image signal power is larger than that of the noise. On the right-hand side of the interception, the noise power is larger than the image signal power. According to what we have learnt from Chapter 2, a perfect denoising can be performed by replacing all singular values of the AWGN corrupted image that lies on the right-hand side of the interception with zero, which corresponds to the hard thresholding process in Equation 4.2. The reconstructed image will be the best-denoised image with respect to Equation 4.1. In other words, we have provided an intuitive prove on the optimality of Equation 4.1, and its solution method in Equation 4.4. The question to completely solve the denoising problem is to find the interception point. There are a number of methods proposed in literature to determine the best low-rank approximation to f, where the threshold t is selected to separate the singular values between Part I (from $\mathscr{s}_1$ to $\mathscr{s}_s$) and Part II (from $\mathscr{s}_{s+1}$ to $\mathscr{s}_M$), and thus will satisfy

$$\mathscr{s}_1 \geqslant \mathscr{s}_2 \geqslant \cdots \geqslant \mathscr{s}_s > t \geqslant \mathscr{s}_{s+1} \geqslant \cdots \geqslant \mathscr{s}_M. \tag{4.6}$$

Let us be reminded that the decreasing eigenvalues of the noise signal in Figure 4.1(c) is the result of the sorted eigenvalues with random distribution around a constant value, which is the power of the noise signal. In other words, the threshold value should be selected according to

$$\mathscr{s}_s > t = 2\sqrt{M}\sigma_\eta \geqslant \mathscr{s}_{s+1}, \tag{4.7}$$

where M is added because

$$\sigma_\eta^2 = \frac{\sum_{m=1}^{M}\sum_{n=1}^{M}(\eta(m,n) - \mu_\eta)^2}{M \times M}. \tag{4.8}$$

Since AWGN has $\mu_\eta = 0$, therefore, Equation 4.8 can be rewritten as

$$\sigma_\eta^2 = \frac{\sum_{m=1}^{M}\sum_{n=1}^{M}\eta(m,n)^2}{M \times M}, \tag{4.9}$$

$$\begin{aligned} M^2\sigma_\eta^2 &= \sum_{m=1}^{M}\sum_{n=1}^{M}\eta(m,n)^2 \\ &= \sum_{\ell=1}^{M}\mathscr{s}_{\eta,\ell}^2 = M\mathscr{s}_\eta^2, \end{aligned} \tag{4.10}$$

$$M\sigma_\eta^2 = \mathscr{s}_\eta^2. \tag{4.11}$$

Noted that Equation 4.11 makes use of $\mathscr{s}_\eta = \mathscr{s}_{\eta,1} = \mathscr{s}_{\eta,2} = \cdots = \mathscr{s}_{\eta,M}$ in theory, even though in reality all M singular values of the noise image varies up and down from $\mathscr{s}_\eta$. The "2" in Equation 4.7 is due to the fact that the threshold should be chosen to separate the eigenvalues of the image that have higher power than that of the noise. Since all the eigenvalues of the noisy image $\mathscr{s}_\ell = \mathscr{s}_{\upsilon,\ell} + \mathscr{s}_\eta$ for $\ell = 1, \ldots, M$, where $\mathscr{s}_{r,\ell}$ is the eigenvalue of the noise-free image. As a result ℓ should be chosen to have $\mathscr{s}_{\upsilon,\ell} \geqslant \mathscr{s}_\eta$, which in turn requires $\mathscr{s}_\ell \geqslant 2\mathscr{s}_\eta > \mathscr{s}_{\ell+1}$. Therefore, we shall choose t according to the following.

$$\begin{aligned} \mathscr{s}_s &\geqslant t = 2 \cdot \mathscr{s}_\eta, \\ &\geqslant 2\sqrt{M}\sigma_\eta. \end{aligned} \tag{4.12}$$

This is an intelligent guess of the optimal threshold value. It is vivid that the approximation in Equation 4.11 will not yield the optimal t. Some rigorous investigations have been conducted to search for the optimal threshold value, where $t = 2.02\sqrt{M}\sigma_\eta$ was proposed in [12]. This threshold has only 1% difference with that derived in Equation 4.13. Another work in [25] is based on an asymptotic framework yields the optimal threshold value given by

$$t = \frac{4}{\sqrt{3}}\sqrt{M}\sigma_\eta. \tag{4.13}$$

All these are correct with respect to a set of prescribed asymptotic framework. In the following we shall keep using t given by Equation 4.12, which is also known as the optimal singular value shrinker [25]. Before we show you the denoising performance of singular value hard thresholding with the optimal singular value shrinker, we shall take a turn to investigate the estimation of σ_η which is required to compute t.

4.2.1 Noise Estimation

According to Equation 4.12, the singular values of the noise image are all smaller than $2\sqrt{M}\sigma_\eta$. As a result, we can conjecture that the singular values of the noise-corrupted image with value smaller than $2\sqrt{M}\sigma_\eta$ are noise components. Therefore, we can estimate the noise image with the following SVD.

$$\eta = U\begin{bmatrix} 0 & 0 \\ 0 & \begin{bmatrix} \mathfrak{s}_{s+1} & 0 & \cdots\cdots & 0 \\ 0 & \mathfrak{s}_{s+2} & 0 \;\cdots & 0 \\ \vdots & \ddots & \ddots\;\ddots & \vdots \\ 0 & \cdots & \cdots\; 0 & \mathfrak{s}_M \end{bmatrix} \end{bmatrix} V^T. \tag{4.14}$$

Once we have the noise image, we can compute the noise power by the traditional method

$$\sigma_\eta^2 = \frac{\sum_{m=1}^{M}\sum_{n=1}^{M}\eta(m,n)^2}{M \times M}, \tag{4.15}$$

or the MAD method in Section 1.5.2.2, then the threshold value will be obtained by Equation 1.14. As a result, a recursive trial and error algorithm can be developed for each selection of $s \in [1, M]$. The optimal noise estimation is obtained when a particular $s + 1$ will yield a noise variance estimation σ_η obtained from either Equations 4.15 or 1.14 where the corresponding t given by Equation 4.12 satisfies Equation 4.6. Showing in Listing 4.2.1 is the MATLAB implementation of the σ_η estimation using Equation 4.15.

Listing 4.2.1: Estimating σ_η by Equation 4.15.

```
function [k,sigma] = svdsigmaest(f)
  [i,j]=size(f);
  m=min(i,j);
  sm = sqrt(m);
  [u s v] = svd(f,0);
  for k = 1:m-1
    a=s;
    a(1:k,1:k)=zeros(k,k);
    noise=u*a*v';
    n = reshape(noise,1,[]);
    sigma=std(n);
    t =2*sm*sigma;
    if ((s(k,k)>=t)&&(t>s(k+1,k+1))) break;
  end
end
```

In a similar manner, the MATLAB function that estimates σ_η by Equation 1.14 is given by Listing 4.2.2.

Listing 4.2.2: Estimating σ_η by Equation 1.14.

```
function [k,sigma] = svdsigmamad(f)
  [i,j]=size(f);
  m=min(i,j);
  sm = sqrt(m);
  [u s v] = svd(f,0);
  for k = 1:m-1
    a=s;
    a(1:k,1:k)=zeros(k,k);
    noise=u*a*v';
    sigma=derisigmaest(noise);
    t =2*sm*sigma;
    if ((s(k,k)>=t)&&(t>s(k+1,k+1))) break;
  end
end
```

If we execute both functions with `f` equals to an AWGN corrupted *Sculpture* image with $\sigma_\eta = 50$ by Listing 4.2.3

Listing 4.2.3: Estimating σ_η by `svdsigmaest` and `svdsigmamad`.

```
>> [k0,sigma0]=svdsigmaest(f);
>> [k1,sigma1]=svdsigmamad(f);
```

The $\sigma_\eta = 45.6663$ with `k=50` is estimated by `svdsigmaest`, while $\sigma_\eta = 47.9679$ with `k=39` is estimated by `svdsigmamad`. It can be observed that the MAD-based noise variance estimation can provide better estimation. Furthermore, `svdsigmamad` is also shown to be able to provide better stability in estimation, while in some cases (particular choice of σ_η, or noise-free photos) `svdsigmaest` fails to convert and cannot provide a reasonable σ_η estimation. This observation coincides with our discussion in Section 1.5.2.2, where the derivative helps to provide better estimation stability.

Please note that when the signal to noise ratio of the image is too low, or when the noisy image contains large structural damage such as heavy salt and pepper noise corruption, the estimation of the above two MATLAB functions might return k with $k + 1$ equals to the image size. At this moment, the estimated σ_η is not reliable, and other methods should be applied to seek a better σ_η estimation.

4.2.2 Denoising Performance

Image denoising by hard thresholding on the singular values can be found in literature back to the '30s [19]. A simple MATLAB implementation is shown in Listing 4.2.4 to help to investigate the performance of the SVD singular value hard threshold denoising method.

Listing 4.2.4: SVD image denoising by hard thresholding.

```
>> [M,N]=size(f);
>> sqrtm = sqrt(M);
>> [u,s,v] = svd(f);
>> sigma=50;
>> ths = 2*sqrtm*sigma;
>> thg = (4/sqrt(3))*sqrtm*sigma;
>> ks = max(find(diag(s)>ths));
>> kg = max(find(diag(s)>thg));
>> gs = u(:,1:ks)*s(1:ks,1:ks)*v(:,1:ks)';
>> gs = imgtrim(gs);
>> gg = u(:,1:kg)*s(1:kg,1:kg)*v(:,1:kg)';
>> gg = imgtrim(gg);
```

The image considered in this simulation is the AWGN corrupted *Sculpture* image with $\sigma_\eta = 50$ (i.e. the source image of Figure 1.8(b)). The PSNR of the AWGN corrupted image is 14.1626 dB. The first part of Listing 4.2.4 performs SVD singular value hard thresholding with known σ_η. The optimal shrinker threshold in Equation 4.12 is applied to give denoised image `gs` as shown in Figure 4.2(a) with PSNR = 19.9404 dB, and the optimal threshold in Equation 4.13 is applied to give denoised image `gg` as shown in Figure 4.2(b) with PSNR = 20.4275 dB. It can be observed that the noise around the metal meshes has been completely alleviated with both thresholds. At the same time, it can be observed that both denoised

(a)

(b)

(c)

Figure 4.2 Denoising result obtained by SVD singular value optimal thresholding on AWGN corrupted *Sculpture* image with $\sigma_\eta = 50$ (noisy image: Figure 1.8(b), PSNR = 14.1626 dB) using (a) shrinker threshold in Equation 4.12 (PSNR = 19.9404 dB), (b) optimal threshold in Equation 4.13 (PSNR = 20.4275 dB), and (c) `svdsigmamad` to estimate σ_η and thresholding singular values (PSNR = 19.5701 dB).

images suffer from a new kind of noise which can be observed from the denoised image with horizontal and vertical lines. In particular, the "back" of the sculpture, and the metal meshes under the bench in Figure 4.2(b) obtained from the optimal threshold estimated by Equation 4.13 are observed to be heavily corrupted with this new kind of noise when compared to that in Figure 4.2(a) obtained from the shrinker threshold estimated by Equation 4.12. This is caused by the discontinuity along the horizontal and vertical image boundaries. However, the objective performance measured by PSNR has shown that the optimal threshold in Equation 4.13 outperforms that of the shrinker threshold in Equation 4.12. To understand these conflicting observations, we shall consider another MATLAB simulation listed in Listing 4.2.5.

Listing 4.2.5: SVD image denoising by hard thresholding with σ_η estimated by `svdsigmamad`.

```
>> [M,N]=size(f);
>> [u,s,v] = svd(f);
>> [k,sigma]=svdsigmamad(f);
>> ge = u(:,1:k)*s(1:k,1:k)*v(:,1:k)';
>> ge = imgtrim(ge);
```

The denoising result obtained from Listing 4.2.5 is shown in Figure 4.2(c) with PSNR = 19.5701 dB, which is lower than both obtained from the shrinker threshold and the optimal threshold. However, it can be observed that the visual quality of the denoised image in Figure 4.2(c) is better than that of (a) and (b). We can thus conjecture that the asymptotic optimality of the threshold obtained from Equation 4.13 can provide the best objective denoise performance. However, it also tends to over truncate the singular values of the denoised image which starts to hurt the visual quality. The shrinker threshold using the real σ_η does not yield a better result than that obtained from Listing 4.2.5 because some of the truncated singular values have heavy contribution from the image than that from the noise, and thus lowering the quality of the resulting denoised image. Listing 4.2.5 uses a threshold that satisfies Equation 4.6, and thus providing the best-denoised image visual quality effect.

4.3 Blocked SVD

With reference to Section 2.3, the SVD hard thresholding denoising method can be improved with block processing, which helps to localize the image power and noise. The best threshold value for the singular values of each block can be estimated with `svdsigmamad` to achieve the best denoising result. Showing in Listing 4.3.1 is the MATLAB function `blocksvd` that implements blockwise operation of `svdhardth` on an input image `f` with block size `L x L`. The output of the function is the denoised image `g`. The function `blocksvd` is implemented in a similar way as that of `blockdct` which makes use of the MATLAB built-in function `blockproc`.

Listing 4.3.1: Block SVD image denoising by hard threshold and estimated σ_η.

```
function g = blocksvd(f,L)
   fun = @(x) svdhardth(x.data);
   g = blockproc(double(f), [L L], fun);
end
```

SVD hard threshold will be applied to each `L x L` image block `f` with the function `svdhardth`. The function `svdhardth` estimates the singular value threshold value using `svdsigmamad`. Furthermore, the output of `svdhardth` is the denoised image block `g` with size `L x L`.

Listing 4.3.2: SVD hard threshold with estimated σ_η.

```
function g = svdhardth(f)
  [u s v]=svd(f,0);
  [k,sigma]=svdsigmamad(f);
  g = u(:,1:k)*s(1:k,1:k)*v(:,1:k)';
  g = imgtrim(g);
end
```

The application of `blocksvd` is shown in Listing 4.3.3 with block size given by the variable `bsize`. In this particular example, the block size `bsize` is determined to divide the image into 32×32 non-overlap blocks.

Listing 4.3.3: 32×32 non-overlap block SVD image denoising by hard threshold and estimated σ_η.

```
>> [m,n]=size(f);
>> bsize=32;
>> g=blocksvd(f,bsize);
```

The denoising result obtained from Listing 4.3.3 is shown in Figure 4.3(a) with PSNR = 19.7760 dB. It can be observed that the blocked SVD helps to localize the image signal power and noise power, thus improving the objective quality of the denoised image, by giving a higher PSNR when compared to that obtained by other SVD methods in this chapter. However, blocking artifacts are observed in the denoised image just, as that discussed in Section 2.3.1. As a result, the same trick of averaging the block SVD denoised image with that obtained from a half-shifted image, as that discussed in Section 2.3.1 is applied to alleviate the blocking artifact.

(a)

(b)

Figure 4.3 Denoising result obtained by block SVD by hard thresholding and estimated σ_η on AWGN corrupted *Sculpture* image with $\sigma_\eta = 50$, where block size of 32×32 and noise estimation by `svdsigmamad` are applied: (a) non-overlap block method (PSNR = 19.7760 dB), and (b) average of two non-overlap block denoised images with one of the images having block half-shifted (PSNR = 20.5629 dB).

Listing 4.3.4: Average of two non-overlap block SVD image denoised results obtained by hard threshold and estimated σ_η, where the block of one image is half-shifted when compared to the other.

```
>> [m,n]=size(f);
>> bsize=32;
>> halfsize=bsize/2;
>> g=blocksvd(f,bsize);
>> fshift=f(halfsize+1:m-halfsize,halfsize+1:n-halfsize);
>> gf=blocksvd(fshift,bsize);
>> g2=double(f);
>> g2(halfsize+1:m-halfsize,halfsize+1:n-halfsize)=gf;
>> gx=(g+g2)/2;
>> g3=imgtrim(uint8(gx));
```

The PSNR of the denoised image obtained in Listing 4.3.4 is 20.5629 dB which is the highest for all the SVD-based image denoising methods discussed in this chapter. Furthermore, the blocking artifacts are alleviated, as it can be observed in Figure 4.3(b). Visually, the image noise content is lowered, and most objects are more well defined, including the metal mesh under the *bench*, and the "*back*" of the sculpture. However, the straight line artifacts are still observed in the denoised image, even though it is not as intense as that observed in other SVD image denoising results.

Previous simulation results have shown that the method can achieve small objective improvement when compared to other methods presented in Chapter 3. The SVD hard threshold method has also shown subjective quality improvement without blurring problem. However, at the same time this method will introduce new artifacts into the denoised image, which in some cases are more annoying than that of AWGN. Besides AWGN, does SVD hard threshold perform well with mixed noise-corrupted image denoising? Showing in Figure 4.4(a) is the denoised image obtained from `svdhardth` with the SVD singular value threshold estimated by `svdsigmamad` for the mixed noise-corrupted *Sculpture* image in Figure 1.11(a), where the PSNR is 13.4352 dB, which is very close to that of the original SAP noise-corrupted image with PSNR of 13.4351 dB. It is vivid that the SAP passes through the denoising algorithm and can be clearly observed in the denoised image. The situation got improved with block SVD hard threshold denoising by `blocksvd`. Figure 4.4(b) shows the denoised results of the same mixed AWGN and SAP noise-corrupted *Sculpture* image obtained by non-overlap block SVD with hard threshold and estimated σ_η using 32×32 block, where the PSNR is 16.3370 dB. It can be observed that some of the SAP got removed because of the localization property of the blocking procedure. The denoising result of the SAP should get improved with decreasing block size which helps to improve the localization property. Figure 4.4(c) is the denoised result with

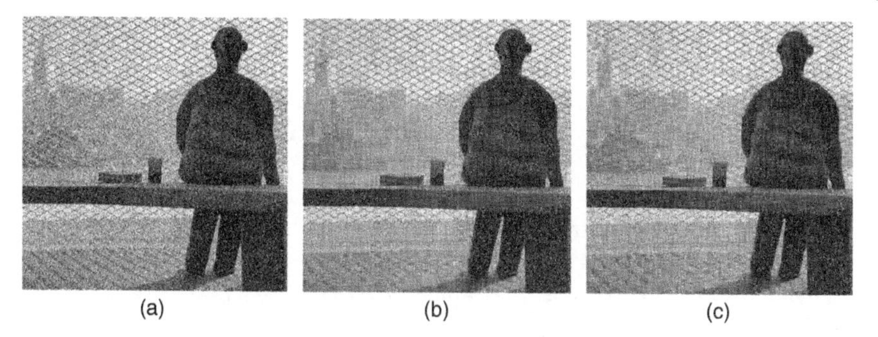

(a) (b) (c)

Figure 4.4 Denoised result obtained by SVD hard thresholding on mixed AWGN with $\sigma_\eta = 50$ and SAP with density of 0.05 corrupted *Sculpture* image (noisy image: Figure 1.11(a), PSNR = 13.4351 dB) using: (a) non-blockwise method with σ_η and singular value threshold estimated by `svdsigmamad` (PSNR = 13.4352 dB), (b) non-overlap blockwise method with block size of 32×32 (PSNR = 16.3370 dB), and (c) non-overlap blockwise method with block size of 16×16 (PSNR = 16.2194 dB).

block size reduced to 16×16, where the PSNR is 16.2194 dB. The SAP is further suppressed and the blocking artifact has been improved, such that the backdrop buildings are more clearly observed. However, decreasing the block size will increase the mixing of the image signal and noise in each block, thus not bringing improvement in the PSNR. Section 4.4 will present methods to help to alleviate the SAP noise-corrupted images.

4.4 The Randomized Algorithm

The SVD is very sensitive to outliers. As a result, it cannot be used to denoise SAP corrupted images. This is because the SAP is generated by Poisson process, which are perfect outliers. To tackle the SAP, we resort to randomized algorithm relying on $\mathcal{L}_1$ regression approach to reach an approximated SVD. We need some freedom to perform the SVD approximation, and in the application of image denoising, the freedom can be obtained by reducing the approximation to be $\mathcal{L}_2$ optimal with respect to the number of singular values to be replaced by zero in the given matrix. This approximation problem is reformulated to be a randomized range finder in [27, 28]. The goal of a randomized range finder is to produce an orthogonal matrix Q with as few columns as possible such that

$$\| \upsilon - QQ'f \| \leqslant \varepsilon, \tag{4.16}$$

for some desired tolerance ε. Q can be discovered by observing that the pairwise distances among a set of points in Euclidean space are roughly maintained when projected onto a lower-dimensional Euclidean space [31]. This basic idea leads to the generation of a Gaussian random matrix P with k columns, and then use f as a linear map for projection. Intuitively, this operation is equivalent to randomly sample the range of f, and the construct an orthonormal basis for these vectors, which will give us the desired Q. In summary, the algorithm to find Q will look like

1. Generate a random matrix P of size $n \times k$, which in MATLAB will be `P=randn(n,k);`
2. Generate an $m \times k$ matrix $Z = f \cdot P$, which in MATLAB will be `Z=f*P`.
3. Generate an orthonormal matrix Q using QR factorization $Z = Q \cdot R$, which can be computed with MATLAB built-in function `qr` as `[Q R] = qr(Z,0)`.
4. The rank k approximation of f will be given by $g = QQ'f$, which in MATLAB will be `g=Q*Q'*f`.

This algorithm takes an input and samples it with sampling parameter k where we can choose k equals to the desired rank of the final approximated matrix. The success of this algorithm depends on the rate of decay of the singular values. If the decay is slow, which usually happens with noisy image with low PSNR or heavily corrupted with SAP noise, the truncated singular values will still contain significant variance in f, and hence failed to denoise the image. To alleviate the effect of the outliers, Q *power iteration* [4] is applied to concentrate energy into the dominant eigenvectors in each iteration, hence reduced the effects of the outliers. The Q power iteration can be implemented in MATLAB as

```
for i = 1:q
    Z=f*(f'*Z);
end
```

Putting them all together, the randomized SVD (RSVD) [27] for rank k approximation will be implemented by the function `rsvd(f,k,q)` which is listed in Listing 4.4.1.

Listing 4.4.1: RSVD.

```
function g=rsvd(f,k,q)
  [m,n]=size(f);
  P=randn(n,k);
  Z=f*P;
  for i = 1:q
    Z=f*(f'*Z);
  end
  [Q R] = qr(Z,0);
  g=Q*Q'*f;
end
```

(a)

(b)

Figure 4.5 Denoised image obtained from RSVD (Listing 4.4.2), where the source image of (a) is the AWGN corrupted *Sculpture* image with $\sigma_\eta = 50$ (Figure 1.8(b)), and the PSNR is 19.4267 dB; and (b) is the mixed noise-corrupted *Sculpture* image in Figure 1.11(a), where the PSNR is 18.0147 dB.

Assume that we have already estimated k by `svdsigmamad`, or some other methods, or even by making an intuitive guess (such as one-fifth of M, the matrix size), and set $q = 2$ to alleviate the outlier effect, `rsvd` can be invoked by MATLAB Listing 4.4.2 to denoise images, where the input parameters are set to $k = 39$ (which is the rank estimated by `svdsigmamad` for AWGN corrupted *Sculpture* image), and $q = 2$ (2 is chosen to ensure a good enough energy concentration).

Listing 4.4.2: RSVD image denoising.

```
g=rsvd(f,39,2);
```

Showing in Figure 4.5 are the denoised image obtained from Listing 4.4.2, where the source image of Figure 4.5(a) is the AWGN corrupted *Sculpture* image with $\sigma_\eta = 50$ (source image: Figure 1.8(b)), and that of (b) is the mixed noise-corrupted *Sculpture* image (source image: Figure 1.11(a)). The PSNR of the two denoised images are 19.4267 and 18.0147 dB, respectively. It can be observed that most of the AWGN are alleviated in both denoised images, and in particular the SAP has been eliminated in Figure 4.5(b). However, when we compare Figure 4.5(a) and (b), it is vivid that the residual noise in (b) is higher than that of (a). Section 4.4.1 will discuss an iterative adjustment method to improve the performance of `rsvd` for the SAP corrupted images.

4.4.1 Iterative Adjustment

The basic idea of iterative adjustment is to add a weighted denoised image of each iteration back to the input noisy image as the input of the next iteration, such that at iteration $\ell + 1$, the input to the RSVD denoise algorithm is given by $u^{\ell+1}$.

$$u^{\ell+1} = (0.5 - w)g^{\ell} + (0.5 + w)u^{\ell}, \tag{4.17}$$

where g^{ℓ} and u^{ℓ} are the denoised image and input noisy image at the ℓ-th iteration, respectively, with $w \in [0, 0.5]$ is the weight assigned to the image components at the ℓth iterations to generate the input image at the $(\ell + 1)$th iteration. Finally, the input image is initiated with $u^1 = f$. To further improve the efficiency of the iterative adjustment, the weight w will be made adaptive with respect to the results obtained at each iteration. To demonstrate how that can be done, we shall present a simple weight adaptation technique, adapting the weight with respect to the iteration number, such that the larger the iteration, the lower the weight. Because when the weight is low, the input image generated for the next round will be close to the statistical mean image between u^{ℓ} and g^{ℓ}, this will generate the optimal input image $u^{\ell+1}$ when g has been optimally denoised, and thus the input and output images will be similar to each other, such that $g^{\ell} \approx u^{\ell}$, and hence the weight $w = 0$, and the combination in Equation 4.17 will be equivalent to statistical means. The MATLAB implementation of the above discussed iterative adjustment algorithm is shown Listing 4.4.3.

Listing 4.4.3: Iterative RSVD.

```
u=f;
for i=1:500
  g=rsvd(u,39,2);
  u=((0.5-exp(-1*i))*g+(0.5+exp(-1*i))*u);
end
```

Other techniques discussed in Chapters 2 and 3 can also be added to the above MATLAB source to make it more efficient, such as using the rate of change of the image denoised image power in successive iteration as the stopping criteria instead of a fixed 500 runs. Such improvement will be the exercise for our readers by the end of this chapter.

When the AWGN corrupted *Sculpture* image with $\sigma_\eta = 50$ and the mixed noise-corrupted *Sculpture* image are applied to Listing 4.4.3, the resulting denoised images are shown in Figure 4.6(a) and (b), respectively, with PSNR equal to 19.7876 and 18.2989 dB, respectively. It is vivid that the objective denoising performance of the iterative adjustment implementation of the RSVD is better than that obtained by other algorithms considered in Sections 4.2 and 4.3 of this chapter. In addition, although the PSNR between these two denoised images are not the same, the two images are observed to be very similar. No SAP are observed in either denoised images. Most of the AWGN are alleviated in the two images, although the denoised image in Figure 4.6(b) is observed to have a

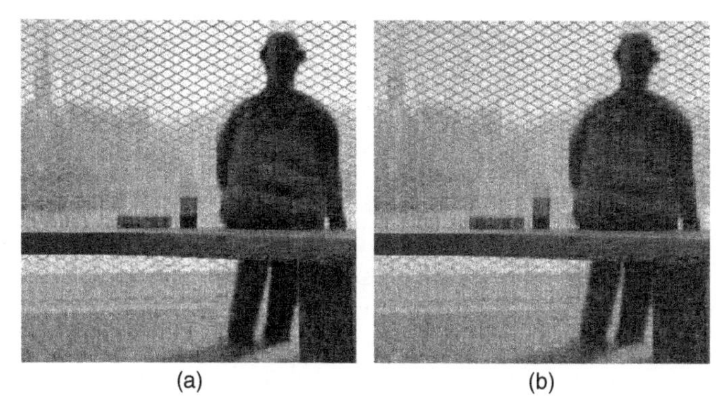

(a) (b)

Figure 4.6 Denoised image obtained from iterative RSVD (Listing 4.4.3), where the source image of (a) is the AWGN corrupted *Sculpture* image with $\sigma_\eta = 50$, and the PSNR is 19.7876 dB; (b) is the mixed noise-corrupted *Sculpture* image in Figure 1.11(a), where the PSNR is 18.2989 dB.

little more noise residual than that in Figure 4.6(a). Nevertheless, most objects in both denoised images have strong and clear edges. The two denoised images suffer from very few blurring effects. To further improve the denoised result, noise localization can help. In other words, block RSVD should be applied to the iterative adjustment. Furthermore, the block RSVD should have overlapped blocks to take care of the blocking artifacts. Again, these are direct applications of methods discussed in Sections 4.2–4.4 of this chapter, and thus will be left as exercise for the readers of this book.

4.5 Summary

SVD has long been a basic tool for signal processing and analysis. But the SVD is less explored for noise estimation and removal in images. In this chapter, we have firstly shown how to infer the noise level according to image singular values out of SVD, due to the fact that the sparsity property of the image signal in SVD space, while noise does not follow with the sparsity property. In addition, we have discussed how to denoise image by hard thresholding the singular values of the noisy image. SVD hard thresholding has shown to be able to achieve very nice denoising results. However, we should understand that when the image is corrupted by AWGN, it will not only affect limited number of singular values. Instead, it will affect all the singular values. Therefore, hard threshold will never be able to truly denoise the noisy image. Furthermore, not all natural images has low-rank. Some of the images have full-rank. As a result, low-rank approximation by removing part of the eigen-structure of the image will induce noises into the denoised image.

In practice, hard thresholding scheme requires an appropriate selection of threshold. How to determine the threshold has been discussed in this chapter. However, the reader should understand that it has been shown that these low-rank recovery schemes are sub-optimal in the matrix denoising problem, and in general, they may not provide a good estimate of the underlying image matrix to be recovered. To improve the SVD hard threshold denoising performance, block processing, which is a signal localization method has been presented in this chapter. The blocking artifacts induced by block processing can be altered by overlapping blocks. Finally, we discussed the application of random SVD low-rank image matrix approximation to achieve SAP corrupted image recovery. We have also discussed the application of iterative adjustment for performance improvement, which can apply to any of the SVD hard thresholding method presented in this chapter.

SVD-based image processing method is robust and usually has good performance. It will be interesting for our readers to keep on working with SVD-based method. Chapter 5 will take a detour from traditional linear transformation-based methods to other more advance image denoise methods.

Exercises

4.1 Modify MATLAB function listed in Listing 4.4.3 from a fixed 500 runs to the more intelligent stop criteria that depends on the rate of change of the MSE between successive iterations. Given an additional parameter ε, and g_{k-1}, g_k and g_{k+1} as the denoised image obtained at iteration $k-1$, k, and $k+1$, respectively, the iterative improvement loop should break when $\frac{mse(g_k, g_{k-1})}{mse(g_{k+1}, g_k)} < \varepsilon$.

4.2 Create a new denoise method that applying iterative RSVD to overlapped blocks of the noisy image, where the noisy image is divided into nine overlapped blocks with 50% overlaps between adjacent blocks. Comment on the PSNR of the denoised image obtained from your new function, and what is the cause of the performance deviated from that obtained from iterative RSVD.

4.3 Create a function `g=irvsd(f,k,q,i,epsilon)` with reference to the MATLAB code in Listing 4.4.3, where `f` is the noisy image, `k` is the estimated singular values threshold, `q` is the iteration power, `i` is the maximum number of iterations, `epsilon` is the PSNR difference in between iterations, and `g` is the denoised image. The function will provide a denoised image `g` with minimized PSNR.

4.4 The low-rank SVD approximation has the problem of inducing artifacts that aligns with the horizontal and vertical axes. This is because the image boundary is also an important features of the image, and thus resulting the artifacts. Noted that the high frequency subbands ("HL" and "LH" subbands) of 2D wavelet decomposition show strong horizontal, and vertical information of the image. Therefore, we should be able to use the directional features of SVD to remove the noise in these subband images. Develop a function that perform the following

1. Perform 3-level decomposition of the noisy image using "db1" wavelet.
2. Perform hard thresholding on the "HH" subband images with scale shrink threshold values as that in Listing 3.5.6 will be applied.
3. Perform SVD image denoised by hard thresholding as that in Listing 4.2.5 on the "LH" and "HL" subband images.
4. Reconstruct the denoised image.

Apply the developed function to denoise an AWGN corrupted *Sculpture* image with $\sigma_\eta = 25$, and report both the PSNR and the resulting denoised image. Comment on the performance, both objective and subjective performances, of the wavelet SVD denoise algorithm performance when compared to the traditional SVD in Listing 4.2.5 performance.

4.5 It has been discussed in Section 4.2.2 that the SVD threshold denoising method will suffer from a new kind of noise that observed as multiple vertical and horizontal lines aligned with image borders. This is caused by the discontinuity along the horizontal and vertical image boundaries. In this example, we shall investigate if such structural noise can be alleviated. With reference to Figure 4.7,

1. Corrupt the *Sculpture* image by AWGN with $\sigma_\eta = 25$ to generate a noisy image f as shown in Figure 4.7(a).
2. Perform SVD hard thresholding by `svdsigmamad` to denoise the noisy image f using the same code as that in MATLAB Listing 4.2.5 to generate the denoised image `gf`. Report the PSNR and print the image.
3. Derive a MATLAB script that performs symmetric extension on f to generate Figure 4.7(b).
4. Derive a MATLAB script to rotate Figure 4.7(b), and then extract a square image of size 725×725 at the center of the image to obtain Figure 4.7(c).
5. Perform SVD hard thresholding by `svdsigmamad` to denoise the image as shown in Figure 4.7(c) using the same code as that in MATLAB Listing 4.2.5 to generate an intermediate denoised image `gee` as shown at Figure 4.7(d).
6. Perform image rotation with $-45°$ to the denoised image `gee` in Figure 4.7(d) to obtain `ge` in Figure 4.7(e).

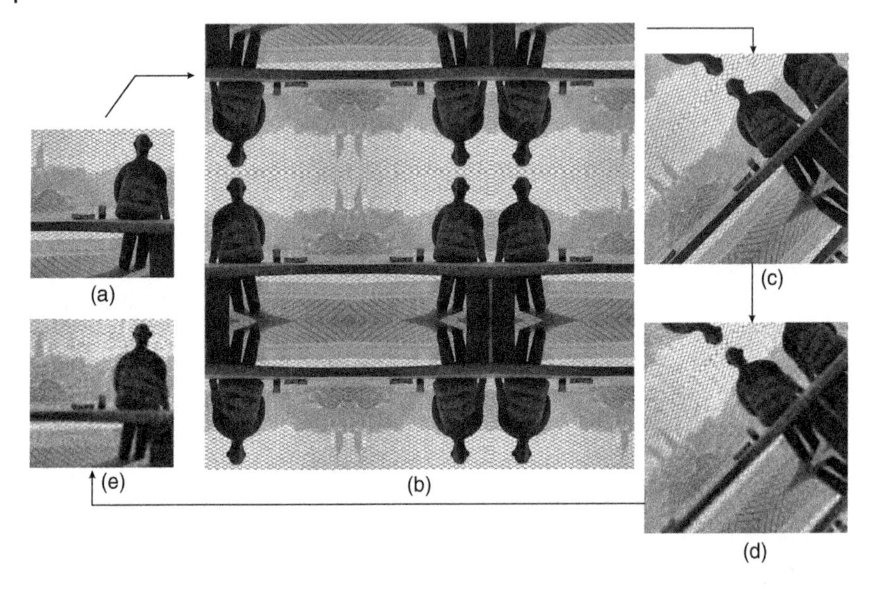

Figure 4.7 (a) AWGN corrupted noisy image f; (b) symmetrical extension patched noisy image; (c) 45° rotated image cropped at the center with size of 725×725 window; (d) denoised image obtained from (c); and (e) -45° rotated image of (d) cropped at the center with size of 512×512.

   (a) Report the PSNR and print the image.
   (b) Does `ge` suffer from the same structural noise as that in `gf`?
7. Generate an average image `g1` from `gf` and `ge`.
   (a) Report the PSNR and print the image.
   (b) Does the image `g1` suffer from the same structural noise as that in `gf`?

5

Variational Method

The Wiener filter in Section 2.2 has shown to be an effective image denoising method. The cons of Wiener filter is the blurring effect observed in the denoised image. A prior information can be applied to Wiener filter to alleviate the blurring problem in the denoised image, such that the denoised image will be obtained as the solution to the following optimization problem.

$$\min_{g} \; \|g - f\|^2 \quad \text{subject to} \quad \lambda V(hg) \leq \epsilon, \tag{5.1}$$

where ϵ is a predefined small number and λ is the weight balance on the function $V(\cdot)$. The first part of Equation 5.1 is similar to the mean squares error (MSE) depicted in Equation 2.9, while the second part of Equation 5.1 introduces an additional cost function $V(hg)$ on the denoised image g, known as regularization function, that imposes some regularity on the optimization solution, i.e. the denoised image. The entry to the regularity function includes a linear operator $h \in \mathbb{R}^{L \times M}$ which is an analysis transformation that preprocesses g before computing the regularization term, with L being an operator-dependent arbitrary number and M is the size of the image. The regularization function helps to convey some prior knowledge about the denoising image, which is independent from any specific information that can be inferred from the noisy image. In this regard, the more carefully the regularity is modeled, the better the quality of the denoised image. Natural images usually exhibit a smooth spatial behavior, except around some locations (such as object edges), where discontinuities arise. Therefore, the quality of the denoise result obtained from solving Equation 5.1 will strongly depend on the ability of the linear operator h and the regularization function V to model such specific type of regularities. Nevertheless, the optimization problem in Equation 5.1 resembles the constrained rank minimization problem in Chapter 4.1, with the rank of the image replaced by a regularization function. Among the sophisticated regularization models developed in the field of image denoising, the most popular one will be discussed in a sequel together with a specific choice of h.

Digital Image Denoising in MATLAB, First Edition. Chi-Wah Kok and Wing-Shan Tam.

Companion website: www.wiley.com/go/kokDeNoise

5.1 Total Variation

One of the models developed for the regularization function is the *total variation* (TV). The total variation model relies on the flow of gradient descent of the image smoothness. It aims to smooth the image as much as possible, that is the chosen norm between adjacent pixels is small, while it has to be as non-smooth as possible at the image edges. The total variation model was first proposed in [46] and is defined as

$$V(g) = \sum_{m=1}^{M} \sum_{n=1}^{N} \sqrt{(g[m+1,n] - g[m,n])^2 + (g[m,n+1] - g[m,n])^2}. \tag{5.2}$$

This model is isotropic and not differentiable. To make it easy to track mathematically, some algorithms use the following anisotropic version of the total variation model

$$V(g) = \sum_{m=1}^{M} \sum_{n=1}^{N} |g[m+1,n] - g[m,n]| + |g[m,n+1] - g[m,n]|. \tag{5.3}$$

The image denoising optimization problem in Equation 5.1 is modified as

$$\min_{g} \mathscr{P} = \min_{g} \{V(g) + \lambda |g - f|^2\}, \tag{5.4}$$

where the optimization function $\mathscr{P}$ is the total variation model $V(g)$ summed with the Lagrange multiplier λ of the sum of squares difference of the denoised image g and the noisy image f. The following will consider different solution methods to solve the above image denoising optimization problem with the isotropic total variation model.

5.1.1 Rudin–Osher–Fatemi (ROF) Model

The idea of applying total variation method in image denoising is first proposed by Rudin et al. in 1992 [46], which is known as Rudin–Osher–Fatemi (ROF) model. The denoising method is applied to AWGN corrupted image in first place, but later extended to apply to images corrupted with different noise sources. The ROF model is also known as a kind of total variation regularization, which aims at removing excessive noise or spurious details (rapid intensity changes not arose from original image features) from the noise-corrupted image, which is usually presented in the form of high total variation (i.e. the large integral of the absolute gradient of the signal). Therefore, it is possible to reduce the noise from the image without harming the original important image features, by reducing the total variation of the signal subject to a particular cost function, and in general the MSE is chosen as the cost function. Therefore, the total variation optimization problem in Equation 5.4 can be rewritten as

$$\int_{\Omega} V(g)dmdn + \frac{\lambda}{2}\int_{\Omega}(g-f)^2 dmdn, \tag{5.5}$$

where Ω is the collection of pixel locations $[m, n]$ that covers the given image f. The total variation model $V(g)$ in Equation 5.2 is chosen for Equation 5.5.

5.1.2 Le-Chartrand-Asaki (LCA) Model

Later in 2007, Le, Chartrand and Asaki proposed a modified ROF model to denoise Poisson noises (e.g. SAP) corrupted images [33], known as the Le–Chartrand–Asaki (LCA) model. The LCA model is also a total variation regularization model, but the dependence of the noise with respect to image features (e.g. edges) is considered in the optimization process. It is greatly suitable to denoise Poisson noise corrupted image rather than AWGN corrupted one. The LCA model is given by

$$\int_{\Omega} V(g)dmdn + \beta\int_{\Omega}(g - f \cdot \log g)dmdn, \tag{5.6}$$

where the data fidelity term is introduced (the second term in Equation 5.6 controlled by the parameter β) with $\beta > 0$ being a positive definite number, aiming at minimizing the difference between the noise-free image and the noisy image. It should be noted that the first term of Equation 5.6 is the same as that presented in the ROF model, where V_g is given by Equation 5.2. Based on the assumption that Ω is a bounded open subset of $\mathbb{R}^2$ with Lipschitz boundary, f is a positive and bounded function in Ω. Since $\log(g) \in \mathscr{L}^1(\Omega)$ with g being positive everywhere in Ω, a unique solution of the regularization must exist.

5.1.3 Aubert-Aujol (AA) Model

The above models are developed based on the additive noise model, and they are effective in removing additive noise. For the multiplicative noise, e.g. gamma noise, it will require another stream of TV regularization models to solve the problem. Aubert and Aujol proposed a modified ROF model [2] dedicated for removing gamma noise, known as the Aubert–Aujol (AA) model, which is given by

$$\int_{\Omega} V(g)dmdn + \beta^{-1}\int_{\Omega}\left(\log g + \frac{f}{g}\right) dmdn, \tag{5.7}$$

where the first term is the same total variation regularization term applied in the ROF model as depicted in Equation 5.2. The second term is the fitting term with β being the weighted parameter. The optimal denoised image g will yield the division between f and g be the mean shifted noise, where a unique solution exists. However, this way to model the multiplicative noise will yield instability at low-intensity region of the image. In the following, we shall concentrate on the discussion of the ROF model which helps to denoise AWGN corrupted image.

5.2 Gradient Descent ROF TV Algorithm

The total variation function $V(g)$ in the ROF algorithm is selected to be the gradient of g, and hence the name "total variation." The gradient of g is ∇g which is defined as

$$\nabla g = \left[\frac{\partial g}{\partial m}, \frac{\partial g}{\partial n}\right]^T = \left[g_m, g_n\right]^T, \tag{5.8}$$

where g_m and g_n are defined as the corresponding partial derivative to simplify the following derivations. Thus the norm $|\nabla g[m,n]|$ is given by

$$|\nabla g| = \sqrt{\left(\frac{\partial g}{\partial m}\right)^2 + \left(\frac{\partial g}{\partial n}\right)^2} = (g_m^2 + g_n^2)^{1/2}. \tag{5.9}$$

Noted that $|\nabla g|$ will be large around image edges. As a result, the optimization in Equation 5.5 will yield a $g[m,n]$ that is close to $f[m,n]$, and hence preserved the edges in the denoised image. At image region with low variation, which are the noise-corrupted regions, $|\nabla g|$ will be small, and the optimization will have the freedom to find a $g[m,n]$ that minimizes $|\nabla g|$ while maintaining certain difference between g and f. It can also be observed that this freedom is adjustable by the value of λ. In other words, we can adjust the edge preservation performance by adjusting λ. The larger the λ, the better the edge preservation performance. However, it will also result in large residual noise. Therefore, an appropriate choice of λ is required to make the algorithm to provide the best denoised image.

The optimization function $\mathscr{P}$ in Equation 5.4 is given by

$$\begin{aligned}\mathscr{P} &= \frac{\lambda}{2}(g-f)^2 + |\nabla g| \\ &= \frac{\lambda}{2}(g-f)^2 + \sqrt{\left(\frac{\partial g}{\partial m}\right)^2 + \left(\frac{\partial g}{\partial n}\right)^2}.\end{aligned} \tag{5.10}$$

The minimization of $\mathscr{P}$ will have to satisfy

$$\begin{aligned}\frac{\partial \mathscr{P}}{\partial g} &= 0 \\ &= \lambda(g-f) + \frac{\partial |\nabla g|}{\partial g}.\end{aligned} \tag{5.11}$$

The partial derivative of $|\nabla g|$ can be separated into two dimensions as

$$\frac{\partial |\nabla g|}{\partial g} = \frac{\partial}{\partial m}\frac{\partial |\nabla g|}{\partial g_m} + \frac{\partial}{\partial n}\frac{\partial |\nabla g|}{\partial g_n}. \tag{5.12}$$

The two inner derivatives at the right-hand side of Equation 5.12 are given by

$$\frac{\partial |\nabla g|}{\partial g_m} = g_m(g_m^2 + g_n^2)^{-1/2}, \tag{5.13}$$

$$\frac{\partial |\nabla g|}{\partial g_n} = g_n(g_m^2 + g_n^2)^{-1/2}. \tag{5.14}$$

As a result, the two differential factors in Equation 5.12 can be obtained by considering the partial differentiation of Equations 5.13 and 5.14, which are given by

$$\frac{\partial}{\partial m}\frac{\partial |\nabla g|}{\partial g_m} = g_{mm}(g_m^2 + g_n^2)^{-1/2} - g_m(g_m^2 + g_n^2)^{-3/2}(g_m g_{mm} + g_n g_{mn}), \tag{5.15}$$

$$\frac{\partial}{\partial n}\frac{\partial |\nabla g|}{\partial g_n} = g_{nn}(g_m^2 + g_n^2)^{-1/2} - g_n(g_m^2 + g_n^2)^{-3/2}(g_m g_{mn} + g_n g_{nn}). \tag{5.16}$$

Adding the above two equations together will yield Equation 5.12 as

$$\begin{aligned}\frac{\partial}{\partial m}\frac{\partial |\nabla g|}{\partial g_m} + \frac{\partial}{\partial n}\frac{\partial |\nabla g|}{\partial g_n} &= \frac{(g_{mm} + g_{nn})(g_m^2 + g_n^2) - (g_m^2 g_{mm} + 2g_m g_n g_{mn} + g_n^2 g_{nn})}{(g_m^2 + g_n^2)^{3/2}} \\ &= \frac{g_m^2 g_{nn} + g_n^2 g_{mm} - 2g_m g_n g_{mn}}{(g_m^2 + g_n^2)^{3/2}}.\end{aligned} \tag{5.17}$$

The gradient descent solution of Equation 5.12 will be given by

$$\frac{\partial \mathcal{P}}{\partial g} = \lambda(g - f) + \frac{g_m^2 g_{nn} + g_n^2 g_{mm} - 2g_m g_n g_{mn}}{(g_m^2 + g_n^2)^{3/2}}. \tag{5.18}$$

All of the above partial differential components can be computed by finite difference method.

5.2.1 Finite Difference Method

A partial differential component can be computed by finite difference method in three different ways

1) Forward difference that computes the derivative of a function g is given by

$$\frac{\partial g[i,j]}{\partial m} = \frac{g[i+\delta,j] - g[i,j]}{\delta}. \tag{5.19}$$

2) Backward difference that computes the derivative of a function g is given by

$$\frac{\partial g[i,j]}{\partial m} = \frac{g[i,j] - g[i-\delta,j]}{\delta}. \tag{5.20}$$

3) Central difference that computes the derivative of a function g is given by

$$\frac{\partial g[i,j]}{\partial m} = \frac{g[i+\delta,j]-g[i-\delta,j]}{2\delta}, \tag{5.21}$$

where δ is a non-zero fixed value approaching zero. This book has adopted the central difference method, because it is a non-biased difference method. The high order difference is given by

$$\frac{\partial^2 g[i,j]}{\partial m^2} = \frac{\frac{g[i+\delta,j]-g[i,j]}{\delta} - \frac{g[i,j]-g[i-\delta,j]}{\delta}}{\delta^2}. \tag{5.22}$$

The mixed differences is given by

$$\begin{aligned}\frac{\partial^2 g[i,j]}{\partial m \partial n} &= \frac{\frac{g[i+\delta_m,j+\delta_n]-g[i-\delta_m,j+\delta_j]}{2\delta_m}}{2\delta_n} - \frac{\frac{g[i+\delta_m,j-\delta_n]-g[i-\delta_m,j-\delta_n]}{2\delta_m}}{2\delta_n} \\ &= \frac{g[i+\delta_m,j+\delta_n]-g[i-\delta_m,j+\delta_n]}{4\delta_m\delta_n} - \frac{g[i+\delta_m,j-\delta_n]-g[i-\delta_m,j-\delta_n]}{4\delta_m\delta_n}.\end{aligned} \tag{5.23}$$

We shall choose $\delta = 1$ (hence δ_m, δ_n) which helps to simplify the above finite difference formulation, and hence we obtain the following finite difference computation model for various partial derivative of $g[i,j]$.

$$\left.\begin{aligned} g_m[i,j] &= (g[i+1,j]-g[i-1,j])/2, \\ g_n[i,j] &= (g[i,j+1]-g[i,j-1])/2, \\ g_{mm}[i,j] &= g[i+1,j]-2g[i,j]+g[i-1,j], \\ g_{nn}[i,j] &= g[i,j+1]-2g[i,j]+g[i,j-1], \\ g_{mn}[i,j] &= \frac{g[i+1,j+1]-g[i-1,j+1]-g[i+1,j-1]+g[i-1,j-1]}{4}. \end{aligned}\right\} \tag{5.24}$$

The following MATLAB source code implement the above equation sets for all points in the image.

```
gm  = (g([2:m m],:)-g([1 1:m-1],:))/2;
gn  = (g(:,[2:n n])-g(:,[1 1:n-1]))/2;
gmm = g([2:m m],:)+g([1 1:m-1],:)-2*g;
gnn = g(:,[2:n n])+g(:,[1 1:n-1])-2*g;
gmn = (g([2:m m],[2:n n])+g([1 1:m-1],[1 1:n-1])-g([1 1:m-1],
      [2:n n])+g([2:m m],[1 1:n-1]))/4;
```

Using the above finite difference implementation of various components in Equation 5.17, we can implement Equation 5.17 as

```
num = gnn.*(epsilon+gm.^2)-2*gn.*gm.*gmn+gmm.*(epsilon+gn.^2);
den = (epilson+gm.^2+gn.^2).^(3/2);
delta = num./den;
```

where a small number `epsilon` is added in the above MATLAB source to avoid the unintended consequence of either factors within the computation is zero value.

The parameter `delta` is the computed number for Equation 5.17. Now if we further introduce the concept of time derivative, such that

$$\frac{\partial g}{\partial t} = \frac{\partial \mathscr{P}}{\partial g}, \tag{5.25}$$

where ∂t is the gradient step size between successive approximation of the denoised image g, which is also known as the time variable in the gradient descent algorithm. The discrete version of Equation 5.25 can be written as

$$\frac{(g[i,j]^{k+1} - g[i,j]^k)}{\Delta t} = \lambda(g[i,j]^k - f[i,j]) + \left(\nabla \cdot \left(\frac{\nabla g[i,j]^k}{|\nabla g[i,j]^k|}\right)\right), \tag{5.26}$$

$$g[i,j]^{k+1} = g[i,j]^k + \Delta t \left(\lambda(g[i,j]^k - f[i,j]) + \left(\nabla \cdot \left(\frac{\nabla g[i,j]^k}{|\nabla g[i,j]^k|}\right)\right)\right), \tag{5.27}$$

where g^k and g^{k+1} are the denoised images obtained at the kth and $(k+1)$th iterations of the gradient descent algorithm. The MATLAB implementation of the above equation is given by

```
g = g+t*(delta+lambda.*(f-g));
```

The above iteration imposes the boundary condition of

$$\begin{array}{lll} g^k(0,j) = g^k(1,j), & g^k(M,j) = g^k(M-1,j), & g^k(i,0) = g^k(i,1), \\ g^k(i,N) = g^k(i,N-1), & g^k(0,0) = g^k(1,1), & g^k(0,N) = g^k(1,N-1), \\ g^k(M,0) = g^k(M-1,1), & g^k(M,N) = g^k(M-1,N-1). & \end{array} \tag{5.28}$$

Besides the boundary condition, the gradient algorithm should be initialized with $g^0 = f$ to enable the gradient descent algorithm to iterate to compute $g^1[i,j]$ from $g^0[i,j]$ for all pixel $[i,j]$ in the image using Equation 5.27, and continue with the next k value until it reaches a predefined maximum number of iterations. The following MATLAB Listing 5.2.1 implements this gradient descent total variation denoising method.

Listing 5.2.1: ROF total variation image denoising.

```
function g=roftv(f,itermax,t,epsilon,lambda)

[m,n]=size(f);
g=double(f);
for i=1:itermax
  gm  = (g([2:m m],:)-g([1 1:m-1],:))/2;
  gn  = (g(:,[2:n n])-g(:,[1 1:n-1]))/2;
  gmm = g([2:m m],:)+g([1 1:m-1],:)-2*g;
  gnn = g(:,[2:n n])+g(:,[1 1:n-1])-2*g;
  gmn = (g([2:m m],[2:n n])+g([1 1:m-1],[1 1:n-1])-g([1 1:m-1],
        [2:n n])+g([2:m m],[1 1:n-1]))/4;
```

```
        % Gradient
        num = gnn.*(epsilon+gm.^2)-2*gn.*gm.*gmn+gmm.*(epsilon+gn.^2);
        den = (epsilon+gm.^2+gn.^2).^(3/2);
        delta = num./den;
        %solution
        g = g+t*(delta+lambda.*(f-g));
        % boundaries
        for i=2:m-1
              g(i,1)=g(i,2);
              g(i,n)=g(i,n-1);
        end
        for j=2:n-1
              g(1,j)=g(2,j);
              g(m,j)=g(m-1,j);
        end
        g(1,1)=g(2,2);
        g(1,n)=g(2,n-1);
        g(m,1)=g(m-1,2);
        g(m,n)=g(m-1,n-1);
        g=imgtrim(g);
    end
    end
```

We can apply `roftv` to denoise image by the following MATLAB function call

Listing 5.2.2: ROF total variation image denoising simulation.

```
>> g=roftv(f,500,0.2,1,0.01);
>> g=roftv(f,500,0.2,1,0.05);
>> g=roftv(f,500,0.2,1,0.1);
```

where the maximum iteration is set to be `itermax=500`, and the step size is set to be `t=0.2`, which is a small step, and we hope that this small step will provide us better denoise results. While the first run has set `epsilon=1`. Use a large `epsilon` will enhance the effect of the regularization. We shall discuss its effect in the denoised image in a sequel. There are three sets of simulations, each with a different `lambda=0.01, 0.05, 0.1`. The simulation results are plotted in Figure 5.1.

When λ is small, the denoised image will no longer bounded to be close to the noise image in least squares sense. As a result, the regularization will take over the control in the denoising result, and thus resulting in an over smoothed image with most of the objects have weak edges and visually observed to be washed out in the denoised image as shown in Figure 5.1(a). With λ set to be 0.05 and 0.1, the descent denoise results are obtained having PSNR at 18.4671 and 17.7932 dB, respectively. The denoised images are plotted in Figure 5.1(b) and (c), respectively. It can be observed that the objects with both weak and strong edges in the original noise-free image are now have their edges preserved in the denoised images.

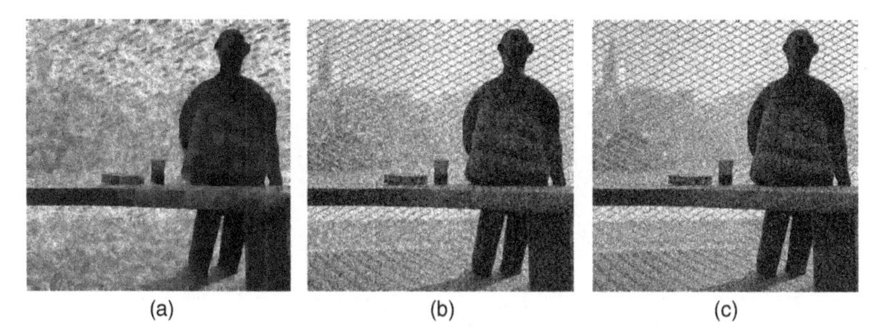

Figure 5.1 ROF TV denoised image using MATLAB Listing 5.2.2 with various λ, where the noisy image is the AWGN corrupted *Sculpture* image with $\sigma_\eta = 50$: (a) $\lambda = 0.01$ ($PSNR = 17.2720$ dB); (b) $\lambda = 0.05$ ($PSNR = 18.4671$ dB); and (c) $\lambda = 0.1$ ($PSNR = 17.7932$ dB).

However, it should be noted that when λ is small, as in the case of Figure 5.1(b), large residual noise will be observed in the homogeneous area. This is due to the noise in homogenous area being preserved by the algorithm when λ is small.

Finally, it should be easy to observe from Figure 5.1(b) that there are artifact staircase noise in the homogeneous area, such as the *Sculpture* back and the metal wire mesh at the lower left hand corner of the image. The cause of this staircase noise will be discussed in Section 5.3.

5.3 Staircase Noise Artifacts

The total variation denoise method has better denoising property with the object edges better preserved. The object outlines are retained in the denoised image, in particular around the highrise buildings at the background. However, watermark alike artifacts are also observed in the denoised image, and the homogeneous area are observed to contain staircase like stepping noise. As an example, the staircase noise observed in Figure 5.1(b), makes the denoised image very unnatural.

The appearance of staircase noise observed in the homogenous area of the denoised image can be investigated by considering the 1D signal f in Figure 5.2(a). The AWGN corrupted signal can be observed in Figure 5.2(b). Noted that when the input signal contains entries have the property of $f[n] \leq f[n+1]$ or $f[n] \geq f[n+1]$, the denoised signal obtained from the optimization in Equation 5.1 must also satisfy $g[n] \leq g[n+1]$ or $g[n] \geq g[n+1]$. If at a particular signal region of f where the signal has the property of $f[n+2] > f[n] > f[n+1] > f[n-1]$ as shown in Figure 5.2(b). Such signal property occurs in the homogeneous region with ramp up intensity which is corrupted by AWGN. In order to satisfy the optimization in Equation 5.1, the

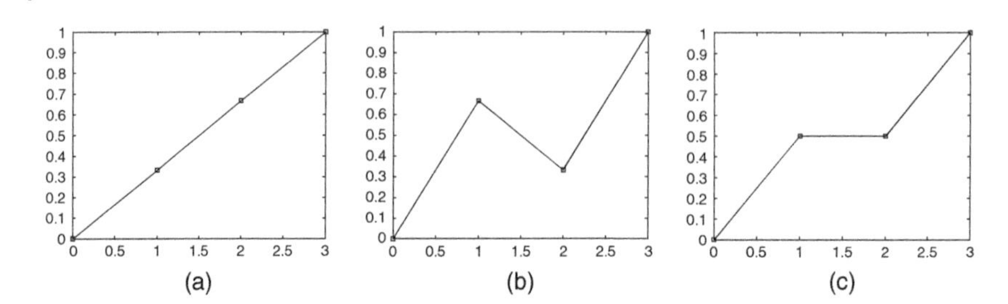

Figure 5.2 Staircase noise artifact: (a) noise-free 1D ramp signal; (b) noise-corrupted signal in (a); and (c) TV denoised signal.

denoised signal will look like Figure 5.2(c), where the denoised signal will have to be $g[n-1] < g[n] = g[n+1] < g[n+2]$. It is this signal intensity property that gives us a visual effect of discontinuity, and thus the staircase noise artifacts. Such denoised image property will be observed in the homogeneous regions far from edges disregard the value of λ. The cause of the existence of staircase artifact is the fundamental failure of Equation 5.1. It is also because of this fundamental limitation of the total variation method leads us to look for better image denoising algorithm, which will be discussed in Chapters 6 and 7.

5.4 Summary

This chapter has reformulated the Wiener filter image denoising problem with the help of a regularization function to a total variation optimization problem. A number of total variation regularization functions have been discussed, and in particular, we have shown how to apply the ROF model to construct the regularization function for image denoising. The finite difference has been applied to implement the ROF total variation image denoising algorithm in MATLAB. The reader should note that there exists other solution methods to the total variation denoising problem, such as the Newton-based method in [40], a duality-based iterative method in [9], a method based on graph-cuts in [10], and another method based on operator splitting in [12], just to list a few.

The ROF total variation denoising method has shown to be a mathematically tractable method that can produce nice denoise results. However, it have its disadvantages, which include over-smoothing, and in some cases, with the multiplier poorly chosen, a cartoon-like appearance of the denoised image will result. Another disadvantage of the ROF total variation denoising is its unique staircase artifacts. Let us acknowledge that the total variation method has potential and should be studied more to derive a perfect total variation-based image denoise algorithm. Chapter 6 will discuss a patch-based denoise method, and in the following chapter, an image denoising algorithm that makes use of the advantage of randomness.

Exercises

5.1 Modify Listing 5.2.1 to use $\frac{|g^{k+1}-g^{k}|}{|g^{k}|} \leq \gamma$ as the stopping criteria, where γ is one of the input variables in your modified code.

5.2 Derive the formula of $g_n = \frac{\partial g}{\partial n}, g_{n,n} = \frac{\partial^2 g}{\partial n^2}$ using central difference method.

5.3 Compare the subjective and objective performance of the ROF TV denoising method using the function `roftv` on AWGN corrupted, AWGN and SAP mixed noise-corrupted and SAP corrupted *Sculpture* image with different sets of function input parameters.

6

NonLocal Means

Among the traditional spatial domain denoising methods, the most typical one is the spatial filtering denoising algorithm, which filters the noise according to the continuity and smoothness of pixel values in a local area with the size equal to the kernel size of the filter. However, such image denoising methods are less effective when the signal-to-noise ratio is low where the noise contaminated the correlation between pixel values in a local area in great extent. Since the objective of the filter is to smooth the pixel variation in a local area, therefore, the filtered image will suffer from blurring problem. To reduce the blurring artifact in the filtered image, the filtering operation should be modified to consider the structure of the image. This chapter will introduce the self-similarity of the image to allow the modified image filtering method to consider the structural characteristics of the whole image. Natural image contains many similar image patches whose positions scattered around the whole image and provides useful redundant information for image denoising. The sparsity-based image denoising presented in Chapter 4 has shown that natural image has self-similarity and thus can be faithfully described by some kind of low-rank representation. Similar idea has been applied in different image processing applications. The most notable application is the fractal image coding. Fractal image coding considers an image as deterministic fractal objects, such that parts of an image are approximated by different parts of the same image. One of the undesirable artifacts of the fractal image coding is the blocking artifacts, which can be alleviated by mean filtering of overlapped blocks. The block overlap can go to the extreme that only one pixel will be considered in the mean filtering between different blocks. A variant of this idea has led to the *NonLocal Means* (NLM) denoising proposed by Buades et al. [5], which estimates each pixel value by averaging the pixel values among the pixels located in the neighboring image patches that are similar to the image patch containing the pixel under concern. This influential work takes full advantage of the redundant information and stimulates extensive research on image denoising using *NonLocal Self Similarity* (NSS). Many variants of NLM have been developed

Digital Image Denoising in MATLAB, First Edition. Chi-Wah Kok and Wing-Shan Tam.

Companion website: www.wiley.com/go/kokDeNoise

and methods exploiting the NSS have evolved from the spatial domain to the transform domain.

6.1 NonLocal Means

Chapter 2 has shown noise reduction can be achieved by per pixel averaging (e.g. mean filtering and Gaussian filtering) with its neighboring pixels. Figure 6.1(a) illustrates a selected sub-image, which contains the pixel under concern $f[p_0]$ and the neighboring pixels p that are included in this averaging operation. This sub-image is also known as the neighboring pixel window $\mathbb{W}[p_0]$. Each pixel $p \in \mathbb{W}$ will be assigned with a weight obtained from the weighting function $w(\cdot)$ where the entry of the weighting function is the difference between p and p_0. As a result, the weighted average is given by

$$g[p_0] = \sum_{p \in \mathbb{W}[p_0]} w(d(p_0, p))f[p], \tag{6.1}$$

where the function $d(\cdot)$ is a difference function between the two pixels $f[p]$ and $f[p_0]$, which can be defined in various ways depending on the applications. The weighting function $w(\cdot)$ should be normalized, such that the weights among all pixels in the $\mathbb{W}[p_0]$ should be normalized, which implies

$$\sum_{p \in \mathbb{W}[p_0]} w(d(p_0, p)) = 1. \tag{6.2}$$

As an example, when the average is performed equally among all the neighboring pixels, the difference function $d(\cdot)$ and the weighting function $w(\cdot)$ have all

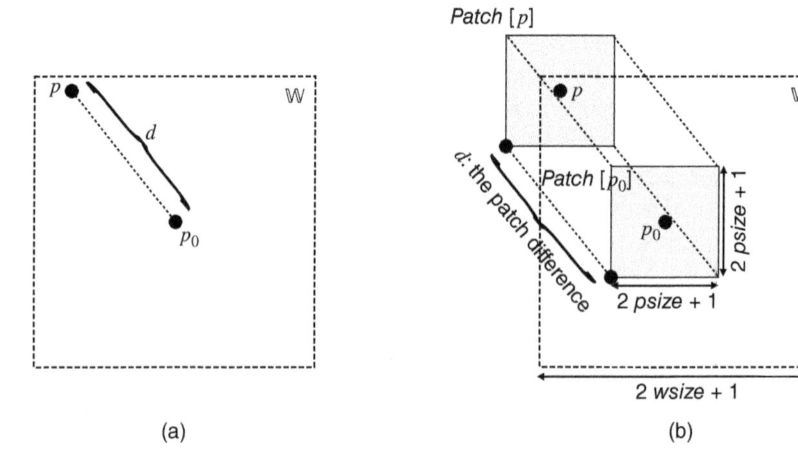

Figure 6.1 Distance function d in (a) Gaussian filtering with distance between pixels $[p_0]$ and $[p]$ is considered; and (b) NLM filtering with mean squares difference between patch $[p_0]$ and patch $[p]$.

their elements identical, such denoising method is known as the mean filtering (Section 2.1). An improvement in the mean filtering is the Gaussian smoothing filtering discussed in Section 2.1.1, where weightings are assigned to neighboring pixels depends on the window $\mathbb{W}$. As another example, it has been shown in Equation 2.8 for the case of Gaussian kernel where the weighting function $w(\cdot)$ will follow a Gaussian function of the distance between the pixel in concern and the neighboring pixels, as shown in Figure 6.1(a). As a result, an intuitive choice for the distance function $d(\cdot)$ will be given by

$$d(p_0, p) = \sqrt{(m_{p_0} - m_p)^2 + (n_{p_0} - n_p)^2}, \tag{6.3}$$

which computes the distance between the two pixels in concern, where m_x and n_x are the row and column locations of pixel x, respectively, and the weighting function is given by

$$w(d(p_0, p)) = \frac{e^{\frac{-d(p_0,p)}{2\sigma_H^2}}}{2\pi\sigma_H^2}, \tag{6.4}$$

where σ_H in the denominator within the exponent is the roll-off factor to mode the distance function to follow a Gaussian function. Therefore σ_H should be chosen to match with the image noise power. The denominator of the overall weighting function should follow the normalization requirement in Equation 6.2.

To further improve the Gaussian filter denoising method, the distance function $d(\cdot)$ in Equation 6.4 can be replaced with the mean squares error between a local block that encloses pixel p_0 with another patch that encloses the neighboring pixel $p \in \mathbb{W}$ in the predefined search window $\mathbb{W}$, as shown in Figure 6.1(b). The weighting function without normalization can be defined similar to that of the Gaussian weighting as

$$w(p_0, p) = \exp\left(\frac{-MSE(patch[p_0], patch[p])}{h^2}\right), \tag{6.5}$$

where the function $patch[p_0]$ extracts a square patch of size $(2 \cdot psize + 1) \times (2 \cdot psize + 1)$ with p_0 as the centre, where *psize* defined by the size of the patch. The search window is defined by the window size *wsize*. These two window parameters should be chosen carefully. Consider *psize*, if it is set to be too large, no similar patch will be found in $\mathbb{W}$. But if *psize* is too small, too many similar patches will be found, which will cause blurring of the denoised image. Common values for *psize* are 2 which yields a 5×5 patch. This patch size is a wise compromise between blurring and texture preservation. A simple MATLAB script `f(m-psize:m+psize,n-psize:n+psize)` (where $p_0 = [\texttt{m}, \texttt{n}]$) can be used to extract the local patch associated with p_0. The `MSE` function in Section 1.8.1 can be invoked to compute the difference between the image patches. Finally, a given parameter h that defines the smoothing factor of this filter will resemble the

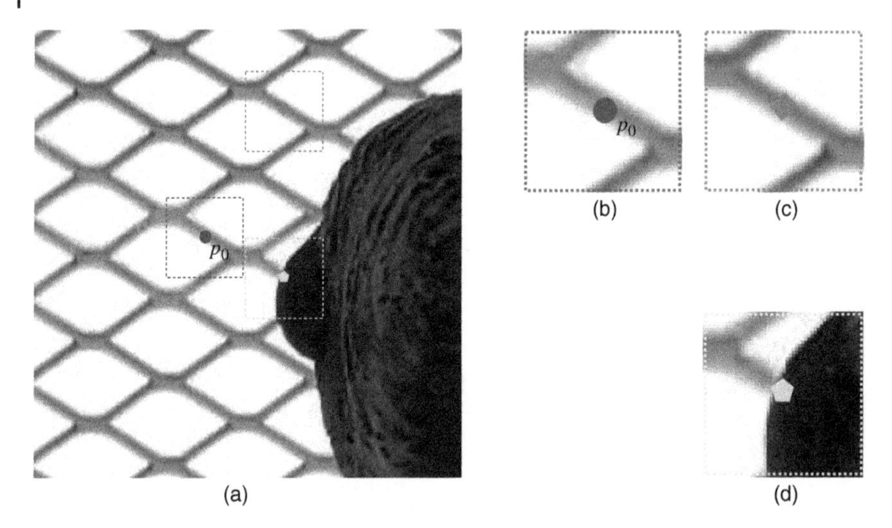

Figure 6.2 Pixel averaging operations: (a) examples of pixels and patches; (b) pixel under concern and the associated patch; (c) neighboring pixel with an associated patch similar to that of the pixel under concern; (d) neighboring pixel with the associated patch being different with that of the pixel under concern.

roll-off factor of the Gaussian filter in Equation 6.4. If h is set too high, the image will become blurry. When an image contains white noise with a standard deviation of σ, h usually takes value in the range of 2σ to 3σ. Figure 6.2 is an example of the computation of the weight. For patch of p_0 is shown in Figure 6.2(b), and assuming Figure 6.2(a) shows the search window $\mathbb{W}$, all the pixels within $\mathbb{W}$ will form their associated local patches, and hence their associated weights will be computed. As you can see, for the patch in Figure 6.2(c), its mean squares difference toward the patch $[p_0]$ shown in Figure 6.2(b) is expected to be very small. Equation 6.5 will yield a weighting value of large magnitude for the patch in Figure 6.2(c) when compared to the patch in Figure 6.2(d). It is vivid that the patch in Figure 6.2(d) looks asimilar to the patch in Figure 6.2(b). Listing 6.1.1 will compute the weightings between the local patch $[p_0]$ with centre pixel $p_0 = [\mathtt{m}, \mathtt{n}]$ and the neighboring patch with centre pixel $p = [\mathtt{p}, \mathtt{q}]$, where the Gaussian weighting are stored at w.

Listing 6.1.1: Windowed pixel difference weighting.

```
patchA = f(m-pszie:m+psize:n-psize:n+psize);
patchB = f(p-pszie:p+psize:q-psize:q+psize);
hs=h*h;
w = exp(-mse(patchA,patchB)/hs);
```

It should be noted that `patchA` is the local patch with centre pixel $p_0 =$ `[m,n]` and `patchB` is the neighboring patch with centre pixel $p =$ `[p,q]`. The variable `hs` in the code denotes h^2 in Equation 6.5 and the last line of the code is the implementation of Equation 6.5.

The weighted pixel value of $f[p]$ can be obtained by `w*f(p,q)`. According to Equation 6.1, the weighted pixel values will have to be summed for all pixels within window $\mathbb{W}$ for the pixel $p_0 = [m, n]$ which defines the region `(m-wsize:m+wsize,n-wsize:n+wsize)` with $(2 \times wsize + 1) \times (2 \times wsize + 1)$ is the size of the window $\mathbb{W}$. Another way to view *wsize* is that it is the radius of the search window. Since the algorithm is quadratic, if the window size *wsize* is the same as the image size $M \times N$, then searching the entire image would be prohibitively expensive with order $\mathcal{O}((MN)^3(psize)^2)$. Because of the inefficiency of taking the weighted average of every pixel for every other pixels in the image, it will be reduced to a weighted average of all pixels in a window that is much smaller than the image size. A natural choice of *wsize* for natural image is 3, which yields a search window size of 7×7, is much more traceable. This particular choice of *wsize* will require a weighted sum of 15^2 pixels instead of $N \times M$ pixels for a $N \times M$ image, and reduced the complexity order to $\mathcal{O}(MN(wsize)^2(psize)^2)$.

The final weighted sum to be presented in the denoised image is given by

$$g[p_0] = \sum_{p \in \mathbb{W}[p_0]} \frac{w(d(p_0, p))}{\text{MAX}(w(d(p_0, p)))} f[p] \tag{6.6}$$

and the MATLAB implementation could be achieved by

```
average=sum(sum(w.*f(m-wsize+offset:m+wsize+offset,n-wsize
    +offset:n+wsize+offset)))
```

It should be noted that the weighting function w is normalized in Equation 6.6, such that the overall weights will not be over 1. We shall need a special treatment in the MATLAB implementation to achieve an accurate normalization, which we will be detailed after we introduce the MATLAB framework for the NLM filtering in Section 6.1.1 for simplicity. Besides the weighting factor, reader should also notice that we need additional pixels in the image boundary. This is because when the NLM filtering is performed over all pixels in the noisy image, there will be chances that the computation kernel (in this case, the patch window) will be partially outside the physical image region, just as we have discussed in Section 1.4.1. As a result, the noisy image should be padded up. In this case, the padding is achieved by `padarray`, with the following MATLAB script, while more details are available in Listing 1.4.1.

```
fpad = padarray(f,[offset,offset], 'symmetric', 'both');
```

where `offset=pszie+wsize`. After the padding, the reader should remember to include the necessary offset `[1+offset,1+offset]` when extracting patches from the `fpad` for `patchA` and `patchB`.

Up to now, we have gone through the required elements for NLM filtering. A MATLAB implementation framework is listed in Listing 6.1.2 based on the above discussions and Listing 6.1.1.

Listing 6.1.2: NonLocal means filtering framwork.

```
hs=h*h;
w=zeros(2*wsize+1,2*wsize+1);
g(m,n)=0;
wmax=0;
 patchA = fpad(m-psize+offset:m+psize+offset,n-psize+offset:
     n+psize+offset);
 for p=-wsize:wsize
 for q=-wsize:wsize
   if (p==0&&q==0) continue; end
   patchB = fpad(m+offset+p-psize:m+offset+p+psize,n+offset
       +q-psize:n+offset+q+psize);
   w(p+wsize+1,q+wsize+1) = exp(-mse(patchA,patchB)/hs);
   wmax = max(wmax,w(p+wsize+1,q+wsize+1));
 end
end
w(wsize+1,wsize+1)=wmax;
w=w/(sum(sum(w)));
g(m,n) = sum(sum(w.*fpad(m-wsize+offset:m+wsize+offset,n-wsize+
    offset:n+wsize+offset)));
```

The normalization of the weighting factor is performed under the factor sum(sum(w)), which includes the sum of all weights assigned to the pixels within $\mathbb{W}[p_0]$. The highest weight wmax is assigned to p_0 which resembles the highest weight in the centre of the Gaussian filter kernel. Such assignment will also prevent pixels within the search window from over-weighting when compared to that assigned to other pixel p within the window $\mathbb{W}[p_0]$.

The complete MATLAB implementation is given in Listing 6.1.3 that implements the NLM filtering denoising algorithm.

Listing 6.1.3: NonLocal means filtering.

```
function g=nlm(f,psize,wsize,h)
offset = psize + wsize;
fpad = double(padarray(f,[offset,offset],'symmetric','both'));
[i,j] = size(f);
g = zeros(i,j);
hs = h*h;
w = zeros(2*wsize+1,2*wsize+1);
for m= 1:i
    for n = 1:j
```

```
            wmax = 0;
            patchA = fpad(m-psize+offset:m+psize+offset,n-psize+offset:
                 n+psize+offset);
            for p=-wsize:wsize
                for q=-wsize:wsize
                    if (p==0 && q==0) continue; end
                    patchB = fpad(m+offset+p-psize:m+offset+p+psize,
                        n+offset+q-psize:n+offset+q+psize);
                    w(p+wsize+1,q+wsize+1) = exp(-mse(patchA,patchB)/hs);
                    wmax = max(wmax,w(p+wsize+1,q+wsize+1));
                end
            end
            w(wsize+1,wsize+1) = wmax;
            w = w/sum(sum(w));
            g(m,n) = sum(sum(w.*fpad(m-wsize+offset:m+wsize+offset,
                n-wsize+offset:n+wsize+offset)));
        end
    end
end
```

To understand the performance of the NLM filter, we shall execute the following MATLAB script.

```
> psize=2; wsize=3; h=50;
> g = imgtrim(nlm(f,psize,wsize,h));
```

where `f` is the AWGN corrupted *Sculpture* image with $\sigma_\eta = 25$. The denoised image is shown in Figure 6.3(a) with PSNR of the denoised image is 23.0491 dB. It can be observed from this figure that the AWGN has been smoothed out, and the noise reduction performance is much better than that of Gaussian filter denoising results. In particular, the edges of the image are sharp. This nice denoising result

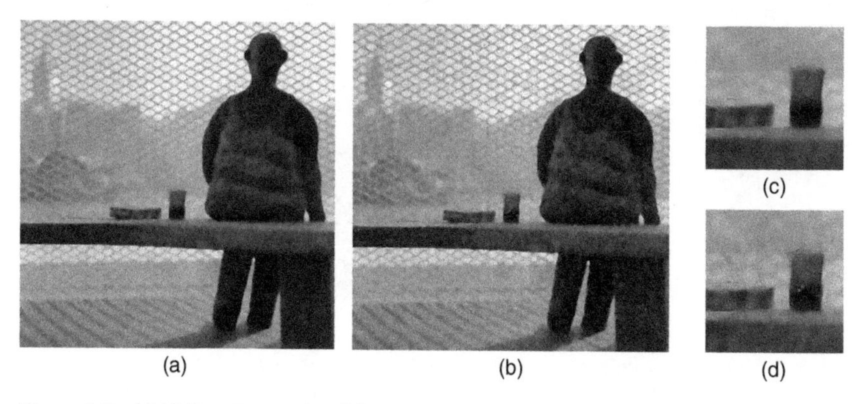

Figure 6.3 NLM filter image denoising results for (a) AWGN corrupted *Sculpture* image with $\sigma_\eta = 25$ (*PSNR* = 23.0491 dB); (b) mixed AWGN with $\sigma_\eta = 25$ and SAP with density of 0.05 corrupted *Sculpture* image (*PSNR* = 22.2157 dB); and (c) and (d) are the zoomed-in portion of the *book and glass* in (a) and (b), respectively.

is obtained because the NLM filter considers the image structure of `patchA` and `patchB` when performing the weighted sum. Different weights will be assigned to each neighboring pixels determined by the similarity between the local block and other patches that containing the neighboring pixels. Similar to the kernel size of Gaussian mean filter, which is now translated to the search range of the neighboring pixels of the local blocks. As a result, the NLM filter not only compares the intensity in a single point but also the geometrical configuration in the sub-image with the neighboring sub-images.

To understand more about the neighboring pixels that are considered in the computation of the weightings in Gaussian filtering and NLM filtering, let us consider the pixel under concern in Figure 6.4(a) (the dot inside the square window near the shoulder of the *Sculpture*). Furthermore, the square window indicates the search window $\mathbb{W}$. As plotted in Figure 6.4(b), the dotted circle encloses the neighboring pixels that are applied in the weighting computation in the Gaussian filtering, where a weighting of greater than half of the peak density of the Gaussian

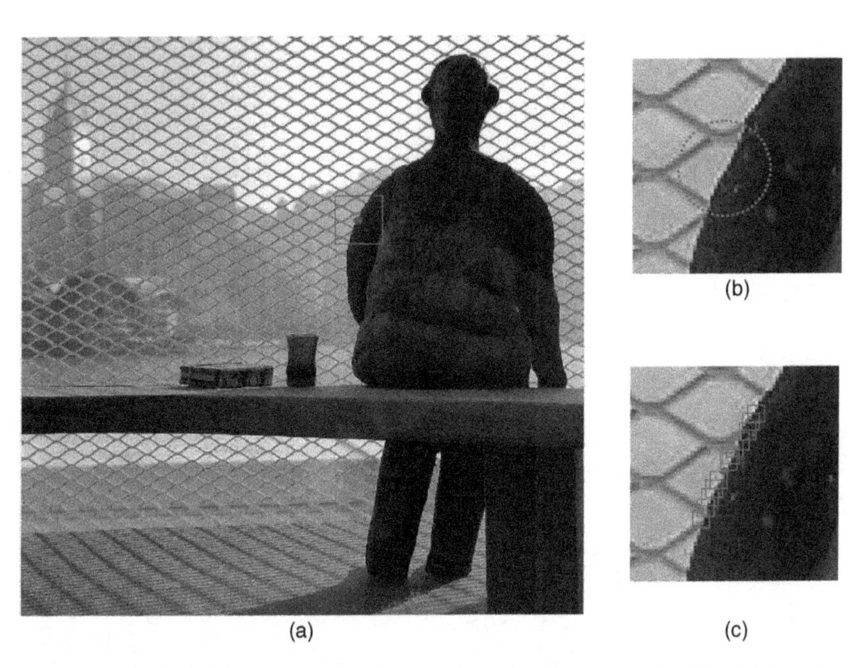

Figure 6.4 Neighboring pixels considered in Gaussian filter and NLM filter: (a) the pixel under concern and the associated search window; (b) pixels involved in Gaussian filtering denoising with weighting factor with magnitude higher than half of the filter peak density; and (c) pixels involved in NLM filtering denoising with weighting factor with magnitude higher than half of the filter peak density (where the boxes are drawn to show the sub-images used to compute the similarity with that of the pixel under concern).

filter is being applied to those pixels. It can be observed that the weights that are higher than half of the peak density of the Gaussian filter forms a circular area disregarding the underlying image structure. Figure 6.4(c) shows the weights higher than half of the peak density of the NLM filter that applied to the image. It can be observed that the pixels with such high weights have similar underlying image structure as that of the pixel under concern. This is a simple evidence to demonstrate the weighted sum of the NLM filter will only strengthen the image structure, and thus producing a denoised image that are not blurred around image edges, as shown in Figure 6.3(a).

When f is replaced with the *Sculpture* image corrupted with the mixed AWGN with $\sigma_\eta = 25$ and SAP with density of 0.05, the denoised image is shown in Figure 6.3(b) with PSNR of the denoised image is 22.2157 dB. It can be observed from Figure 6.3(b) that the AWGN has been smoothed out and with some SAP remained. Figure 6.3(c) and (d) show the zoomed-in images of the *book and glass* region of Figure 6.3(a) and (b), respectively. Unlike other filtering methods, the denoised images obtained by the NLM filter are observed to have strong edges. Unfortunately, the texture region that contains objects with low-intensity difference (i.e. the weak texture region) was observed to have been washed out. The *book and glass* in both Figure 6.3(c) and (d) have shown to be denoised having the sharp edges preserved, while the metal meshes behind them have been washed out because the metal meshes are weak texture. The reason of texture washout problem is simple, neighboring pixels belong to patches that do not match well with the patch of the pixel under concern still got assigned with a weight, where the weighted pixel intensities will be part of the sum that forms the final denoised signal intensity. Moreover, it should be noted that the SAP remains along the metal meshes (short edges) in Figure 6.3(c). These shows the deficiency of the NLM filtering. This problem can be alleviated by hard threshold method to be presented in Section 6.1.1.

6.1.1 Hard Threshold

The NLM filter will always assign weights to neighboring pixels within the search window that depends on the distance between the local patch and the patches associated with pixels inside the searching window. The goal is to applied a weighted sum of neighboring pixels from similar patches to achieve denoising. However, the small weights assigned to neighboring pixels, where the associated patches have large distance from the local patch, have caused the weak texture washout problem in the denoised image. A simple approach to alleviate this problem is to apply hard threshold to the distance between patches. All patches with distance that are far away from the local patch are dissimilar patches, and will be eliminated from the computation of the weighted sum. After eliminating dissimilar patches through hard thresholding, the remaining patches are processed in the weighted sum to generate the denoise pixel intensity with the normalization reapplied to the remaining weights associated with the remaining pixels. The NLM filter with hard threshold `nlmhth(f,psize,wsize,h,th)` is implemented in MATLAB Listing 6.1.4.

Listing 6.1.4: NLM filtering with hard threshold.

```
function g=nlmhth(f,psize,wsize,h,th)
offset = psize + wsize;
fpad = double(padarray(f,[offset,offset],'symmetric','both'));
[i,j] = size(f);
g = zeros(i,j);
hs = h*h;
w = zeros(2*wsize+1,2*wsize+1);
for m= 1:i
    for n = 1:j
        wmax = 0;
        patchA = fpad(m-psize+offset:m+psize+offset,n-psize+offset:
            n+psize+offset);
        for p=-wsize:wsize
            for q=-wsize:wsize
                if (p==0 && q==0) continue; end
                patchB = fpad(m+offset+p-psize:m+offset+p+psize,
                    n+offset+q-psize:n+offset+q+psize);
                w(p+wsize+1,q+wsize+1) = exp(-mse(patchA,patchB)/hs);
                wmax = max(wmax,w(p+wsize+1,q+wsize+1));
            end
        end
        w(wsize+1,wsize+1) = wmax;
        w = w/sum(sum(w));

        for p=-wsize:wsize
            for q=-wsize:wsize
                if (p==0&&q==0) continue; end
                if (w(p+wsize+1,q+wsize+1) <= th)
                    w(p+wsize+1,q+wsize+1)=0;
                end
            end
        end
        g(m,n) = sum(sum(w.*fpad(m-wsize+offset:m+wsize+offset,
            n-wsize+offset:n+wsize+offset)));
    end
end
end
```

The parameters in `nlmhth` are similar to those in `nlm` discussed in Section 6.1, with an added parameter `th`, the hard threshold value, such that the following threshold function will be performed on the normalized weights $w_{p_0}(p)$ according to the value th

$$w_{p_o}(p) = \begin{cases} w_{p_o}(p), & w_{p_o}(p) \leq th, \\ 0, & w_{p_o}(p) > th. \end{cases} \tag{6.7}$$

Noted that the remaining weights associated with the remaining pixels will have to be normalized again to achieve the best and also tractable weighted average sum for each pixel p_0.

Showing in Figure 6.5 are the denoised image obtained by executing the following MATLAB script.

```
> psize=2; wsize=3; h=50; th=0.01;
> g = imgtrim(nlmhth(f,psize,wsize,h,th));
```

where f is the input noisy image. The same set of noisy images that we applied in Figure 6.3 are applied here for evaluating the performance of hard thresholding NLM filtering, where the denoised results are shown in Figure 6.5(a) for the AWGN corrupted image and Figure 6.5(b) for the mixed AWGN and SAP corrupted image, where the corresponding zoomed-in portions for the *book and glass* are shown in Figure 6.5(c) and (d), respectively. The PSNR of Figure 6.5(a) is 22.4341 dB and that of Figure 6.5(b) is 20.9752 dB, which are both lower than those of the simple NLM-filtered counterparts. The degraded objective performance is due to the suppression of far-distance weightings, which reduces the similarity between the matched neighboring patches to that of the local patch under concern. However, it brings the advantage that the washout effect is reduced, where the weak textures are preserved better. The improvement is vivid at the metal meshes in the background when compared to those in the simple NLM filtering (see the outlines of the *book* and the *glass*). The SAP removal has been improved in hard threshold NLM filtering too as the weights for those extreme pixels

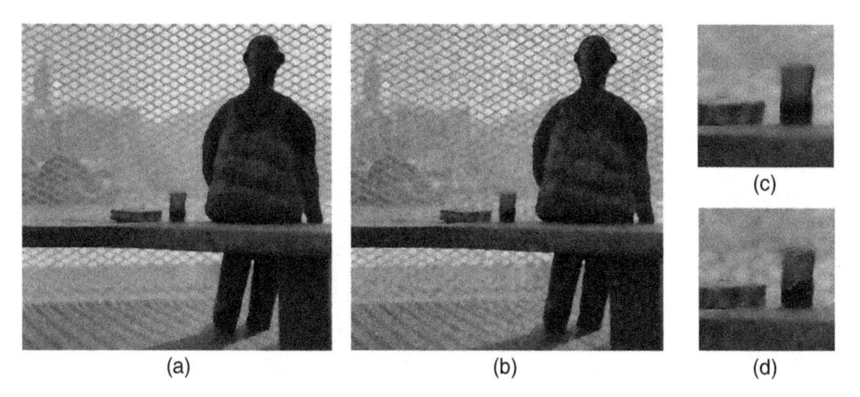

Figure 6.5 Hard thresholding NLM filter image denoising results for (a) AWGN corrupted *Sculpture* image with $\sigma_\eta = 25$ (*PSNR* = 22.4341 dB); and (b) mixed AWGN with $\sigma_\eta = 25$ and SAP with density of 0.05 corrupted *Sculpture* image (*PSNR* = 20.9752 dB); and (c) and (d) are the zoomed-in at the *book and glass* of (a) and (b), respectively.

contaminated by SAP are suppressed. It can be observed that there are less SAP remained in the *glass*. Section 6.2 will present another method that can achieve similar performance improvement when compared to that of the conventional results but at the same time be able to bring in more computational advantages.

6.2 Adaptive Window Size

Please note that there are two windows in NLM filter which are defined by *psize* and *wsize*. Both windows can be made adaptive which means a different *psize* or *wsize* for different pixels during the weighted average operations discussed in Equation 6.5. Before we discuss how can we adapt the window size to achieve better denoised image, we shall first investigate the effect on the denoised image with *psize* and *wsize* changed globally over the whole image. The base of this investigation will be the conventional NLM filter with `psize=2`, `wsize=3`, and `h=50` applying on the *Sculpture* image corrupted with AWGN of σ_η. By executing the following MATLAB script, two figures will be obtained and are presented in Figure 6.6. By observing the trends of the two curves in these two figures, we can derive the best window size adaptation procedure for image denoising by NLM filtering.

```
> psnr1=zeros(1,10);
> wsize=3;
> h=50;
> for psize=1:10
    g=imgtrim(nlm(f,psize,wsize,h));
    psnr1(1,psize)=psnr(g,v);
        end

> psnr2=zeros(1,14);
> psize=2;
> h=50;
> for wsize=5:14
    g=imgtrim(nlm(f,psize,wsize,h));
    psnr2(1,wsize)=psnr(g,v);
        end
```

6.2.1 Patch Window Size Adaptation

It can be observed from Figure 6.6 that as patch size (`psize`) increases, the PSNR of the denoised image decreases. Not only the objective quality of the denoised image decreases, the blurring of the denoised image details such as edges (e.g. the outlines of *book and glass*) and weak texture (e.g. the metal meshes around the *book and glass*) get worst with increasing patch size. This is because the edges and texture are pixels with large intensity difference, which become more robust to the

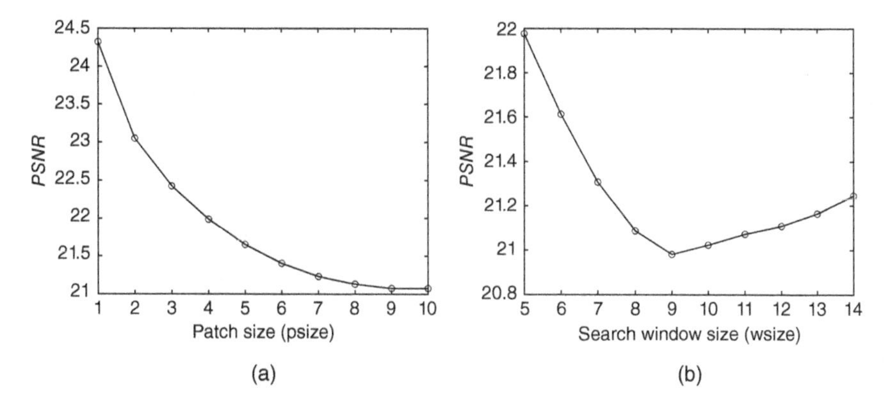

Figure 6.6 The PSNR performance of NLM filtering on AWGN corrupted *Sculpture* image with $\sigma_\eta = 25$ in a relation of (a) changing patch size (`psize`) with `wsize=3`, and (b) changing search window size (`wsize`) with `psize=2`.

weighted sum of the NLM filter. However, homogeneous areas usually have small intensity differences. Therefore, when a large patch window is applied, which is so large that it covers not only the homogenous areas but also cover part of the edge and texture areas, in this case, it results in the *rare patch effect*. This effect describes the patch has lack of redundancy in a search region. In this case, the weighted average will have a much bigger disturbance, and results in higher intensity variation than it should be. In other words, the large patch window size will make the pixels around the homogeneous area to have large intensity variation, and thus blurring the edge and texture, and eventually degrading the subjective quality of the denoised image. Hence, the patch size should adapt to the local texture content of the image patch, which will help to enhance both the subjective and objective quality of the denoised image.

To adapt to the variation of local texture, we shall need a way to find out if individual pixel belongs to texture region or non-texture region. Fortunately, we did that before in Section 1.8.4 with the MATLAB function `tmap`. The `tmap` will return 0 if the pixel is located at homogeneous region, and will return 1 if the pixel is located at edge or texture region. Different patch window size will be assigned to `psize` depending on the result of `tmap`, where the maximum patch window size `pmax` will be adopted when edge and texture region is detected (i.e. output of `tmap` is 1) and the reduced window size `pmax-pratio` will be adopted when homogeneous region is detected (i.e. output of `tmap` is 0). The following MATLAB code implements the assignment of `psize`.

```
psize=pmax-pratio;
if (tmap(m,n)==1)
    psize=pmax;
end
```

The NLM filtering with adaptive patch window size `nlmadp` can be realized by modifying the basic NLM filter function `nlm` by adding the above MATLAB code inside the for loops. It should be noted that the `offset` should be defined by the sum of the maximum patch window size `pmax` and also the search window size `wsize` to avoid any missing pixels at the image boundary. The complete MATLAB code of `nlmadp` is shown in Listing 6.2.1.

Listing 6.2.1: NonLocal means filtering with adaptive patch window size.

```
function g=nlmadp(f,pmax,pratio,wsize,tmap,h)
offset = pmax + wsize;
fpad = double(padarray(f,[offset,offset],'symmetric','both'));
[i,j] = size(f);
g = zeros(i,j);
hs = h*h;
w = zeros(2*wsize+1,2*wsize+1);
for m= 1:i
    for n = 1:j
        wmax = 0;
        psize=pmax-pratio;
        if (tmap(m,n)==1)
            psize=pmax;
        end
        patchA = fpad(m-psize+offset:m+psize+offset,n-psize+offset:
            n+psize+offset);
        for p=-wsize:wsize
            for q=-wsize:wsize
                if (p==0 && q==0) continue; end
                patchB = fpad(m+offset+p-psize:m+offset+p+psize,
                    n+offset+q-psize:n+offset+q+psize);
                w(p+wsize+1,q+wsize+1) = exp(-mse(patchA,patchB)/hs);
                wmax = max(wmax,w(p+wsize+1,q+wsize+1));
            end
        end
        w(wsize+1,wsize+1) = wmax;
        w = w/sum(sum(w));
        g(m,n) = sum(sum(w.*fpad(m-wsize+offset:m+wsize+offset,
            n-wsize+offset:n+wsize+offset)));
    end
end
end
```

Showing in Figure 6.7 is the denoised image obtained by executing the following MATLAB script.

```
> pmax=3; wsize=3; pratio=1; h=50;
> t=tmap(f,50);
> g=nlmadp(f,pmax,pratio,wsize,t,h);
```

Figure 6.7 NLM filtering with adaptive patch size image denoising results for (a) AWGN corrupted *Sculpture* image with $\sigma_\eta = 25$ (*PSNR* = 21.9853 dB); (b) mixed AWGN with $\sigma_\eta = 25$ and SAP with density of 0.05 corrupted *Sculpture* image (*PSNR* = 21.2997 dB); (c) and (d) are the corresponding zoomed-in at the *book and glass* for (a) and (b), respectively.

The same set of corrupted images that have been applied in the evaluation of the basic NLM filtering and NLM filtering with hard thresholding is used here. It can be observed that the PSNR of both images are lower than those of the basic NLM case because the patch window sizes are increased, which is consistent with that presented in Figure 6.6(a). However, the PSNR of the mixed noise image denoised by NLM filtering with adaptive patch size, as shown in Figure 6.7(b) (i.e. PSNR = 21.2997 dB) is found to be higher than that of the NLM filtering with hard thresholding. It is because, smaller patch size is adopted in homogenous area, thus suppressing the SAP in a better way. This can be observed in the zoomed-in images of the *book and glass*, as shown in Figure 6.7(c) and (d) that the metal mesh washout problem have been improved. Moreover, the region inside the metal meshes in both cases are much cleaner and the SAP in the homogenous region of the glass is suppressed. Unfortunately, the strong edges outlining the *book and glass* are blurred, as shown in Figure 6.7(c) and (d), because a larger patch window is adopted.

Other methods have been proposed to modify the NLM filter to adapt the patch window size. Interested readers are referred to [15, 29, 49, 54]. In particular, the NLM filter variant in [15] has proposed to replace the square patch window by arbitrary shapes to take the advantage of the local geometry of the noisy image. The limitation on designing a better patch to be used in NLM filter is limited by our innovation to capture the local geometric features of the image.

6.2.2 Search Window Size Adaptation

The size of search region in NLM algorithm is limited to all pixels in an image due to computational cost. On the other hand, it can be observed from Figure 6.6(b) that large search window size does not always provide a better image denoising result. The reason is similar to that of the patch window size. It can be seen that the size of search window should be small for texture regions, and large for homogeneous regions. Since the number of pixels to be assigned to the weighted average is exactly the same as the search window size. Therefore, when the search window is small, there may not be large enough number of pixels to participate in the weighted average to smooth out the noise, especially when the noise power is high. This is the reason why a large search window should be used for pixel in homogeneous region.

Similarly, when the search window is large, there will be a high chance that a large number of pixels which are not similar to the pixel in the texture region will be participating in the weighted average operation and causes blurring in the denoised image. This is the same observation as discussed in Section 6.1.1. In Section 6.1.1, hard threshold has been applied to limit the number of pixels to participate in the weighted average operation such that it helps to alleviate the edge and texture blurring problem. With the luxury to actually change the number of pixels to participate in the weighted average operation, a small size search window should be applied to pixels in textures and edges.

On the other hand, it can be observed from Figure 6.6(b) that if the search window size is too small, there will not have enough pixels to participate the weighted average denoising operation, and hence there is always a minimal search window size for this method to work with, in order to achieve a good denoising result. In conclusion, there is always a minimal search window size to achieve the best image denoising result. At the same time, the search window size should be made adaptive to the local image characteristics. One of the image features that can be quantified and used to adapt the search window is again the texture map. The idea is similar to that of the NLM filter with adaptive patch size, but we consider the size of the search window instead. Let's consider the maximum search window size is `wsizem`. With `t` being the texture map of `f`, when `t=1` which is the edge and texture region, `wsize=wsizem-wratio` would be applied to reduce the search window size, while when `t=0` which is the homogeneous region, and the maximum window size would be used, such that to improve the denoising result for homogeneous regions. Similarly, the `offset` should be well adjusted to ensure there are sufficient pixels to form a complete search window for pixels around the image boundary. The complete MATLAB code of NLM filtering denoising with adaptive search window size (`nlmadpw`) is shown in Listing 6.2.2.

Listing 6.2.2: NonLocal means filtering with adaptive search window size.

```
function g=nlmadpw(f,psize,wsizem,wratio,tmap,h)
wsize=wsizem;
offset = psize + wsizem;
fpad = double(padarray(f,[offset,offset],'symmetric','both'));
[i,j] = size(f);
g = zeros(i,j);
hs = h*h;
w = zeros(2*wsize+1,2*wsize+1);
for m= 1:i
    for n = 1:j
        wmax = 0;
        if (tmap(m,n)==1)
            wsize=wsizem-wratio;
        end
        patchA = fpad(m-psize+offset:m+psize+offset,n-psize+offset:
            n+psize+offset);
        for p=-wsize:wsize
            for q=-wsize:wsize
                if (p==0 && q==0) continue; end
                patchB = fpad(m+offset+p-psize:m+offset+p+psize,
                    n+offset+q-psize:n+offset+q+psize);
                w(p+wsize+1,q+wsize+1) = exp(-mse(patchA,patchB)/hs);
                wmax = max(wmax,w(p+wsize+1,q+wsize+1));
            end
        end
        w(wsize+1,wsize+1) = wmax;
        w = w/sum(sum(w));
        g(m,n) = sum(sum(w.*fpad(m-wsize+offset:m+wsize+offset,
            n-wsize+offset:n+wsize+offset)));
    end
end
end
```

Showing in Figure 6.8 is the denoised image obtained by executing the following MATLAB script.

```
> psize=2; wsizem=4; wratio=1; h=50;
> t=tmap(f,50);
> g=nlmadpw(f,psize,wsizem,wratio,t,h);
```

Again, the same set of corrupted images that applied in Section 6.2.1 is applied to evaluate the performance of NLM filtering with adaptive search window size, where the PSNR of the two denoised images are lower than that of the basic NLM filtering cases. However, they both have better visual quality than that obtained by the NLM filtering with adaptive patch size. Figure 6.8(a) and (b) show the denoised images. We can observe that the washout problem is further improved, with the noise inside the homogeneous regions more readily removed (cleaner region in between metal meshes and the liquid inside the glass, as

Figure 6.8 NLM filtering with adaptive search window size image denoising results for (a) AWGN corrupted *Sculpture* image with $\sigma_\eta = 25$ ($PSNR = 22.4046$ dB); (b) mixed AWGN with $\sigma_\eta = 25$ and SAP with density of 0.05 corrupted *Sculpture* image ($PSNR =$ 21.8840 dB); (c) and (d) for the corresponding zoomed-in at the *book and glass* for (a) and (b), respectively.

shown in Figure 6.8(c) and (d)). This demonstrates that the rare patch problem can be alleviated by adopting larger search window but smaller patch size for the homogenous region. Moreover, the strong edges (along the *Sculpture* ear and outlines of the objects) are sharper when compared to those in other NLM filtering cases as the search window size is increased and more image information is considered. However, the computational cost is heavily increased.

6.3 Summary

The NLM filter denoising method makes use of the neighborhood similarity to reduce noise in an image. The NLM filter assumes that every small local patch in a natural image has many similar patches in its neighborhood. Image denoising can be achieved by making use of this redundant information to perform weighted average that will help to smooth the noise. The core of the algorithm is to assign a weight to pixel that is estimated from a patch that contains the pixel and another patch that contains the pixel to be denoised. The estimation of this weight should be proportional to the mean squares difference between the two patches. Such an image-denoising algorithm has several issues that limited its applications, which include search window size, patch window size, smoothing parameter, central pixel weight, computational cost etc. This chapter has discussed the methods handling several of these issues and have left a few to be investigated by the readers in the Exercise section. The performance of the NLM algorithm depends on the proper selection of the parameters input to the algorithm. To increase the robustness and also the performance, this chapter has discussed various methods to make these parameters to adapt to the characteristics of the noisy image. The algorithm of NLM filter matches well with characteristics of the natural image and has been widely used in various image-denoising problems, not limited to photos, it also includes autopsy, microscopic images, etc. There are still a lot of potentials in the NLM filtering that can be exploited to produce a better image-denoising algorithm. One of them will be discussed in an exercise in Chapter 7.

Exercises

6.1 Investigate the PSNR by varying the parameter h in Equation 6.5 on denoising the AWGN corrupted *Sculpture* image with $\sigma_\eta = 25$ with $h = 0.5\sigma, \sigma, 2\sigma, 4\sigma, 8\sigma$. [Hints: MATLAB function `nlm` Listing 6.1.3.]

6.2 Mixed NLM filtering

1. Develop the MATLAB function that combines the adaptive NonLocal Means image-denoising algorithm with the hard threshold property as in Section 6.1.1 and the adaptive search window size property as in Section 6.2 into a single image-denoising algorithm.
2. Present the parameter set that the image-denoising algorithm developed in exercise 6.2.1 will provide the best result for the *Sculpture* image corrupted with AWGN at $\sigma_\eta = 50$ in terms of subjective performance and PSNR.
3. Does your program outperform both hard threshold and adaptive search window size algorithm? Can you give us a reason about your observation?
4. What does this observation tell you about the mix and match of algorithms for image denoising?

6.3 Patch kernel

1. Create a new NonLocal Means filter denoising MATLAB function that uses a uniform kernel instead of Gaussian kernel in Equation 6.4 for the computation of distance between two patches. Instead of weighting the mean squares error (MSE) between two patches using a Gaussian kernel, the inverse of the MSE between two patches is directly assigned to be the weight between the associated two pixels. The largest weighting will be assigned to the central pixel within the search window. Of course, all the generated weight will have to go through normalization before it can be used to weight the pixels and summing to generate the denoised pixel. [Hints: remember to find a good use of the smoothing parameter `h` in your algorithm.]

2. Denoise an AWGN corrupted *Sculpture* image using function `nlmhth` (Listing 6.1.4) and compare it to the result using the function developed in Section 6.1. Record and discuss the performance of this image denoising algorithm.

6.4 Centre pixel weighting

1. Applying Stein Unbiased Risk Estimator (SURE) for weight calculation, where the weight of the central pixel in the NonLocal Means filter is replaced by $e^{\frac{-2\sigma_\eta^2 N\times M}{h^2}}$ without modifying the other weights, before normalization. Implement and compare the performance of the NonLocal Means filter using SURE and that of the conventional NonLocal Means filter.
2. In a similar manner, construct the NonLocal Means filter with the central pixel being assigned with a weight of 1 and 0 after the weight normalization, and then renormalize the weight once again (similar to that being done in `nlmhth`).
3. Shows the PSNR obtained by all four types of NonLocal Means filters with different central pixel weights (i:0; ii:1; iii:Max (as in conventional NonLocal Means Filter); iv:SURE) with respect to the *Sculpture* image corrupted by AWGN with $\sigma_\eta = 10, 30, 50$ using `psize=2` and `wsize=3` and assume the noise σ_η is known a prior.
4. Discuss if SURE is a better weight assignment to the central pixel, and let us know analytically why?

7

Random Sampling

When a large amount of mixed Gaussian and salt and pepper noise are added to the *Sculpture* image, as shown in Figure 7.1(a) (same as that shown in Figure 1.10(b)), most of the highrise buildings in the background of the image are barely recognizable and the metal meshes around the *book and glass* have completely washed out because of the image masking effect. When this noise corrupted image is downscaled (lowpass filtered followed by downsampling by a factor of 2) to form a low-resolution image, as shown in Figure 7.1(b), despite the lower in resolution, it appears to contain much less noise and the highrise buildings in the background of the image become more recognizable. To understand the physics behind this observation, let us consider the image spectrum, as shown in Figure 7.1(a) and (b). The solid line shown in Figure 7.1(a) is the spectrum of the noise-free image, and the spectrum of the noisy image is plotted with a dotted line in the same figure. It can be observed that the noisy image has a heavier tail. The sample spectrum is plotted for the image in Figure 7.1(b). It can be observed that the spectrum of the downsampled noisy image and the noise-free image are very similar in every part of the spectrum. This effect can be understood by the fact that the image downsampling operation effectively averages neighboring pixel values and thus lowering the power of the uncorrelated noise among neighboring pixels in the downsampled image. Since the neighboring pixels in natural images are often highly correlated, the downsampling process does not cause much damage to the image information in the downscaled image. As a result, Figure 7.1(b) appears to be less noisy and objects within the image appear to be clearer than that in Figure 7.1(a). Another explanation is that natural images have most energy concentrated in the low spatial frequency regions, whereas the AWGN is uniformly spread over the whole spectrum. Filtering followed by downsampling an image keeps mainly the low-frequency components, which are precisely the components that contain most of the image information. If we try to make use of the lowpass filtering and downsampling

Digital Image Denoising in MATLAB, First Edition. Chi-Wah Kok and Wing-Shan Tam.

Companion website: www.wiley.com/go/kokDeNoise

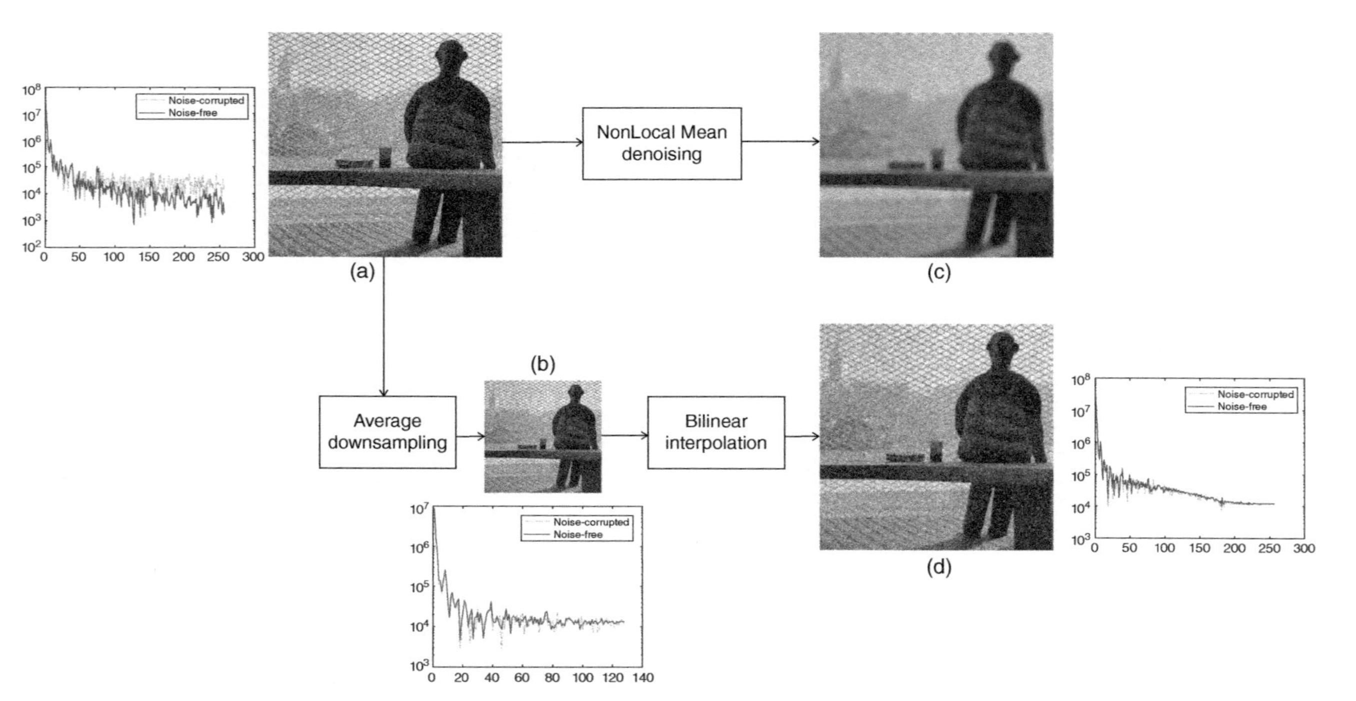

Figure 7.1 Noisy *Sculpture* becomes less noisy by down- and up-sampling, where the associated power spectra is also displayed.

procedure to denoise an image, we can interpolate the downsampled image in Figure 7.1(b) to recover a denoised image, as shown in Figure 7.1(d). The advantage of this denoising method is vivid when compared with the denoised image obtained from the NonLocal Means (NLM) filtering method, as shown in Figure 7.1(c). The NLM filtering method finds it hard to dispose noise close to the perimeters of the image objects, especially when the intensity differences across the object perimeters are low, such as the high-rise buildings in the background of the *Sculpture* image. The NLM filter not only failed to denoise the SAP in the image, the NLM filter amplified the SAP, as shown in Figure 7.1(c), and caused the image to be severely blurred. The nicely denoised image in Figure 7.1(d) is therefore a good demonstration on the denoising capability of the sampling and interpolation method.

However, this method does have a few problems. The number one problem is how to interpolate the low-resolution image to recover a full-size image. The quality of the denoised image depends largely on the interpolation procedure. If denoising is carried out first, it might take away some nice structures that help the interpolation of the low-resolution image. If interpolation is carried out first, the noise will also be interpolated, which makes the pixels that have low signal-to-noise ratio tougher to be denoised. As a result, executing the interpolation and denoising approaches concurrently have the advantages over performing them separately. This will be the topic in Section 7.3. In a sequel, we shall discuss another variant of the sampling and interpolation method to achieve image denoising.

7.1 Averaging Multiple Copies of Noisy Images

To understand how we can derive an efficient image-denoising method with its core formed by sampling and interpolation, we shall have to study how to make use of a prior information of the random nature of the noise to denoise a noisy image. Consider we have K copies of the same image υ corrupted by Gaussian noise that are *independent and identically distributed* (IID) which yields f_ℓ as

$$f_\ell = \upsilon + \eta_\ell, \qquad \ell \in [1, K] \in \mathbb{Z}, \tag{7.1}$$

where η_ℓ is the AWGN with zero mean for the ℓth copy of the noise-corrupted image f_ℓ. Now consider the mean of this set of noisy images.

$$g = \frac{1}{K}\sum_{\ell=1}^{K} f_\ell = \frac{1}{K}\sum_{\ell=1}^{K}(\upsilon + \eta_\ell)$$
$$= \frac{1}{K}\sum_{\ell=1}^{K}\upsilon + \frac{1}{K}\sum_{\ell=1}^{K}\eta_\ell$$
$$= \upsilon + \frac{1}{K}\sum_{\ell=1}^{K}\eta_\ell. \tag{7.2}$$

Since η_ℓ is normally distributed with zero mean and are generated by IID process, it can be readily shown that dividing by $K \gg 1$ will readily make $\frac{1}{K}\sum_\ell^K \eta_\ell \to 0$. The greater K, the closer the noise will sum to zero. Analytically, the noise power in image g is given by

$$\sigma_{g,\eta}^2 = \frac{1}{K}\sigma_\eta^2. \tag{7.3}$$

As a result with $K \to \infty$, g in Equation 7.2 will approach υ,

$$g \xrightarrow{K\to\infty} \upsilon. \tag{7.4}$$

As an example, the following MATLAB will average 10 AWGN corrupted *Sculpture* images with $\sigma_\eta = 50$ (i.e. $K = 10$).

Listing 7.1.1: Multiple noise images averaging.

```
>> v = double(sculpture);
>> f = zeros(size(v));
>> sigma = 50;
>> for i = 1:10
     noise = sigma.*(randn([size(f)]));
     fint = v + noise;
     f = fint + f;
end
>> g=f/10;
>> g=uint8(imgtrim(g));
```

The image obtained from Listing 7.1.1 is shown in Figure 7.2(b). Comparing with one of the noise-corrupted image f_1, as shown in Figure 7.2(a), the image denoising result in Figure 7.2(b) is similar to that in Figure 7.1(d) which is obtained from downsampling and interpolation. This technique of averaging multiple copies of the same scene to obtain a better noise reduced image has been applied in astro-photography, where a large amount of images are captured with telescope to generate an image with extreme details. However, unlike astro-photography, most applications are limited to work with only one image. In this case, we shall have to consider how to generate the rest of the $(K - 1)$ images

(a)

(b)

Figure 7.2 (a) One of the 10 AWGN corrupted *Sculpture* images with $\sigma_\eta = 50$. (b) The average of all 10 AWGN corrupted *Sculpture* images with $\sigma_\eta = 50$ ($PSNR = 24.48$ dB).

from a single source image. A straight forward approach is to corrupt the image with $(K - 1)$ different sets of Gaussian noise. However, such approach does not alter the noise in the first image, and hence statistical average will almost sure have no effect on the noise in the source image. It can be foreseen that the average image will at least contain as much noise as that in the original noisy image.

The above discussion leads to the idea of disruptive effect on the noise component of the original noisy image. There exist a number of methods that can be imposed on the noisy source image to create a disruptive effect on the noise component. A straightforward approach is to corrupt the image with pepper noise which is equivalent to remove the pixels values of a set of selected image pixels. The effect will create an image with missing pixels. A set of images with different sets of missing pixels can be generated, which can then be used to generate a set of Gaussian noise-corrupted images through missing pixel recovery technique, also known as inpainting. These set of recovered Gaussian noise-corrupted images can be averaged to obtain a nice denoised image.

7.2 Missing Pixels and Inpainting

Section 7.1 has discussed two important ideas: first, sampling and reconstruction to achieve image denoising; second, averaging multiple denoised images obtained by the same denoise method with images corrupted by identical and independent noises will improve the denoise result. One of the methods to generate such set of denoised images is to randomly sample the image, and replace the sampled pixel value with zero. These zero valued pixels, also known as missing pixels will be recovered by inpainting. Instead of structured downsampling and then perform conventional image interpolation, random sampling remove pixel values by constructing a collection Ω that contains a set of pixel locations for which $f[p]$ survives the sampling when $[p] \in \Omega$. The recovery of the missing pixels is not obtained through interpolation, but by inpainting. Inpainting corresponds to filling the holes in an image where the holes are the missing pixels. Mathematically, inpainting is related to the matrix completion problem, which aims to pass the partial observations $f[p]$ with $[p] \in \Omega$ of an unknown matrix υ to estimate a matrix g that contains the estimated missing pixels and the partial observations, and g is close to υ. Such a problem is ill posed and completely unsolvable without additional constraints.

Fortunately, the natural image matrix f has low-rank property, as discussed in Chapter 4, and the low-rank property is one of the properties that helps to make the matrix completion problem solvable. Knowing that the image matrix is of low-rank, the matrix completion problem can be formulated as the following optimization problem.

$$\min_{g} \ \text{rank}(g), \qquad \text{s.t.} \quad g[p] = f[p], \qquad \forall [p] \in \Omega, \tag{7.5}$$

where g is the inpainted image. To make the mathematics to be more tractable, we created the following projection function $P_{\Omega}(\cdot)$ to replace the sampling operation in the *subject to* clause in Equation 7.5, such that Equation 7.5 can be rewritten using the projection operator as

$$\min_{g} \ \text{rank}(g), \qquad \text{s.t.} \quad P_{\Omega}(g[p]) = P_{\Omega}(f[p]), \tag{7.6}$$

and the projection function is defined as

$$P_{\Omega}(f[p]) = \begin{cases} f[p], & [p] \in \Omega, \\ 0, & \text{otherwise.} \end{cases} \tag{7.7}$$

For simplicity, we define the following MATLAB operator for the projection operation $P_{\Omega}(\cdot)$

```
ProjO = @(f,Omega) Omega.*f;
```

where `Omega` equals to 1 indicates the corresponding pixels will survive the sampling, otherwise, when `Omega` equals 0, the corresponding pixels will be replaced with 0. It is vivid that the above problem is an NP-hard problem. Furthermore, its complexity grows exponentially with the squares of the matrix dimension. In order to solve the problem in Equation 7.6, convex relaxation can be applied to convert the problem to a convex optimization problem, and this is being done by replacing the rank(g) operation by the nuclear norm $\|\cdot\|_*$ defined in Equation 4.5

$$\min_{g} \|g\|_* \qquad \text{s.t.} \qquad g[p] = f[p], \qquad \forall [p] \in \Omega. \tag{7.8}$$

It has been pointed out in [14] that the solution of Equation 7.8 is equivalent to the solution of Equation 7.5 with a high probability under the condition of strong incoherence. In a sense, this is a tight convex relaxation of the NP-hard rank minimization problem because the nuclear norm sphere is the convex hull of a set of rank one matrices with a spectral norm of 1. More specific theorems and proofs can refer to [7].

7.3 Singular Value Thresholding Inpainting

Singular value threshold (SVT) algorithm is proposed in [7] to solve the nuclear norm minimization problem in Equation 7.8. The SVT algorithm first regularizes the optimization problem in Equation 7.8 as

$$\min_{g} \quad t\|g\|_* + \frac{1}{2}\|g\|_F^2, \qquad \text{s.t.} \qquad P_\Omega(g) = P_\Omega(f), \tag{7.9}$$

where $t > 0$ regulates the contribution of the singular values in the optimization problem. In other words, it is the SVT, and thus, the name of the algorithm. When $t \to +\infty$, the optimization solution will be dominated by the nuclear norm of g and hence will be the optimal solution of Equation 7.8. The solution of the constrained optimization problem in Equation 7.9 can be obtained by the Lagrangian multiplier method as the following

$$L(g, a) = \|g\|_* + \frac{1}{2}\|g\|_F^2 + \langle a, P_\Omega(f - g)\rangle, \tag{7.10}$$

where a is an intermediate matrix having the same dimension as that of g. It has been argued in [7] that if $(\hat{g}, \hat{a})$ are the primal-dual optimal then

$$\sup_a \inf_g L(g, a) = L(\hat{g}, \hat{a}) = \inf_g \sup_a L(g, a). \tag{7.11}$$

Uzawa's algorithm [1] helps to find this optimal pair through the above primal-dual optimal property with an iterative procedure. Starting with $a^0 = 0$, the solution to Equation 7.8 can be obtained when the following iteration converges.

$$\begin{cases} \begin{cases} [U, S, V] = SVD(a^{k-1}), \\ S = \text{sgn}(S) \odot \max(\text{abs}(S) - t, 0), \\ g^k = U \cdot S \cdot V^T. \end{cases} \\ \{\, a^k = a^{k-1} + \delta_k P_\Omega(f - g^k), \end{cases} \tag{7.12}$$

where $\{\delta_k\}$ with $k \geq 1$ and $k \in \mathbb{Z}$ is a sequence of scalar step sizes. The stopping criteria is when $P_\Omega(g^k)$ is close to $P_\Omega(f)$, which can be estimated by $\|P_\Omega(g^k - f)\|_F$ being small enough. However, just as with all kinds of measurement, it should be normalized, and this time, it will be normalized by the Frobenius norm of $P_\Omega(f)$. As a result, the stopping criteria for the SVT algorithm is given by

$$\frac{\|P_\Omega(g^k - f)\|_F}{\|P_\Omega(f)\|_F} \leq \epsilon. \tag{7.13}$$

The following MATLAB function `svt` implemented the iterative solution of Equation 7.12, where the iteration will stop when Equation 7.13 is satisfied which implies the algorithm has converged. The iteration will also stop when the maximum iteration number exceed `maxiter`.

Listing 7.3.1: SVT.

```
function [g,iterations] = svt(f,P,th,delta,maxiter,epsilon)
[m,n] = size(f);
lambda = zeros([m,n]);        % initialization
iterations = 0;
ProjO = @(f,Omega) Omega.*f;
if nargin < 3 th =  sqrt(m*n); end
if nargin < 4 delta = 1; end
if nargin < 5 maxiter = 300; end
if nargin < 6 epsilon = 1e-7; end

for k = 1:maxiter
    [U, S, V] = svd(lambda, 'econ') ;         % update g in Equation 7.12
    S = sign(S) .* max(abs(S) - th, 0) ;
    g = U*S*V' ;
    lambda = lambda + delta* ProjO((f-g),P); % update lambda in eq.(7.12)
    lambda = ProjO(lambda,P) ;
    error = norm(ProjO((f-g),P),'fro' )/norm( ProjO(f,P),'fro' ); %
       termination condition
    if error<epsilon  break; end
    iterations = k ; % update iterations
end
end
```

Figure 7.3(c) shows the SVT algorithm recovered image from a sampled intermediate image of an AWGN corrupted *Sculpture* image with $\sigma_\eta = 25$, as shown in Figure 7.3(a). This intermediate image `fn` is generated by assigning 50% pixels in

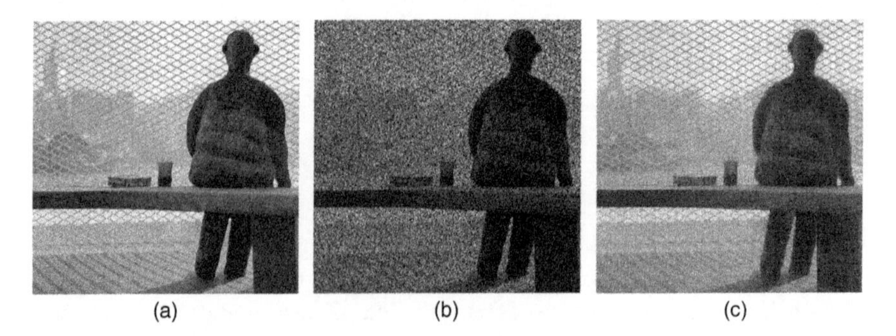

Figure 7.3 Singular value thresholding inpainting: (a) AWGN corrupted *Sculpture* image with $\sigma_\eta = 25$ ($PSNR = 20.6836$ dB), (b) the intermediate image generated from (a) where 50% of pixels in (a) is assigned to zero at random locations ($PSNR = 7.4265$ dB), and (c) singular value threshold algorithm recovered image from (b) ($PSNR = 19.2825$ dB).

Figure 7.3(a) to zero at random locations, as shown in Figure 7.3(b). The following MATLAB script shows the generation of fn.

```
>> [M,N] = size(f);
>> [pos] = sort(randperm(M*N, fix(M*N*0.5)))';
>> MM  = reshape(f(:,:),M*N,1);
>> P = zeros(M*N,1);
>> P(pos) = MM(pos);
>> index1 = find(P);
>> P(index1) = 1;
>> P = reshape(P,M,N);
>> ProjO = @(f,Omega) Omega.*f;
>> fn = ProjO(f,P);
```

The built-in function randperm generates a row vector containing a random permutation of $0.5 \times M \times N$ integers selected randomly in $[1, M \times N]$. Please note that the temporary variable MM is created to ensure only non-zero pixel values will be considered by the projection operator P. However, the sample rate is guaranteed by MM in the above script. Furthermore the sample rate is chosen to be 50% in the above MATLAB script, which means that the intensity of 50% of the pixels will be replaced by 0. The peak signal-to-noise ratio (PSNR) of the noise-corrupted image, f, in Figure 7.3(a) is 20.6836 dB, and that of, fn, in Figure 7.3(b) is 7.4265 dB. The random sampled image can be recovered by using the svt function with the following MATLAB script.

```
>> th=50000; delta=1; epsilon=1e-7; k=500;
>> [g,iterations]=svt(fn,P,th,delta,k,epsilon);
```

where P is the projector that containing the pixel locations of the set Ω, and delta is the step size, which we have chosen to be 1. The SVT th is set to be 50000, and the epsilon that determines the convergence of the algorithm is set at 1e-7. The recovered image is shown in Figure 7.3(c). The PSNR of the recovered image is 19.2825 dB. It can be observed that the image quality in Figure 7.3(c) is not as good as that in Figure 7.3(a), and even the objective quality measure, PSNR, has told us already. However, when we consider the new image $(f + g)/2$ which is shown in Figure 7.4(a). The PSNR of this image is 21.4026 dB which is higher than that of f and fn. It can be conjectured that when 10 individual images recovered from Figure 7.3(a) with 50% of the pixel values at random locations being replaced by 0 values, and the random locations of these 10 images has less than 50% correlation by the SVT algorithm are averaged together with the noisy image f will achieve a higher PSNR and low visual noise quality. This image is shown in Figure 7.4(b), and the PSNR is 22.4795 dB. Both the objective and subjective qualities of the denoised image in Figure 7.4(b) is much better than that of the noisy image f. The details of the MATLAB code is left as an exercise for our reader.

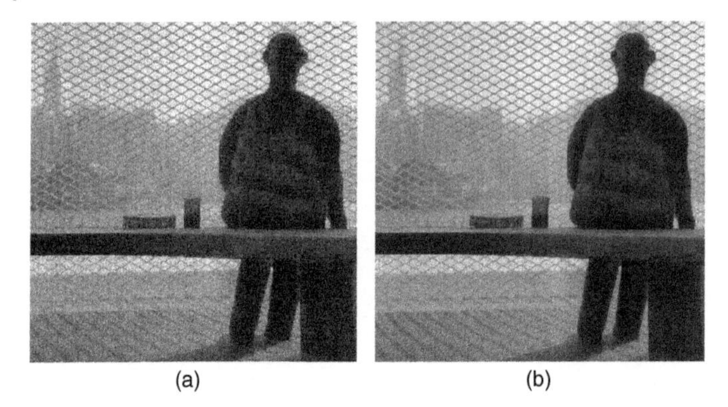

(a) (b)

Figure 7.4 Averaging of multiple denoised image: (a) by averaging images of Figure 7.3(a) and (c) ($PSNR = 21.4026$ dB), and (b) by averaging images of Figure 7.3(a) and 10 singular value threshold recovered images from Figure 7.3(a) with 50% of the pixel values at random locations being replaced by 0 values, and the random locations of these 10 images has less than 50% correlation ($PSNR = 22.4795$ dB).

7.4 Wavelet Image Fusion

Can we further improve the quality of the denoised image obtained from averaging multiple SVT recovered random sampled noisy images? There are two possible angles to answer this question. One is to apply a better image recovery algorithm instead of the SVT. There are numerous possibility in literature, and therefore, we shall leave it as an exercise to the reader. Another approach is to implement a better multiple image fusion method instead of using the average method to obtain the denoised image from multiple recovered images. It is the objective of this section to consider the application of wavelet image fusion that we have discussed in Section 3.8 to combine all the SVT recovered images to a single denoised image. In particular, we shall apply Garrote thresholding in the wavelet image fusion to suppress the Gaussian noise residual in the SVT recovered images. The same function `waveletfus` presented in Listing 3.8.2 will be applied to achieve better denoised image.

Let us consider the image fusion result obtained from fusing Figure 7.3(a) and (b), which is shown in Figure 7.5(a). The image is obtained from the following MATLAB script and the PSNR of this image is 21.0811 dB.

```
>> [M,N] = size(s1);
>> ssize = M*N;
>> [w,j,th] = waveletfus(f,s1,ssize,0,3);
```

where `s1` is the SVT recovered image, as shown in Figure 7.3(c). The output image `w` is the fused denoised image, where the objective image quality is definitely better, but there are good and bad about the subjective image quality. It can be observed

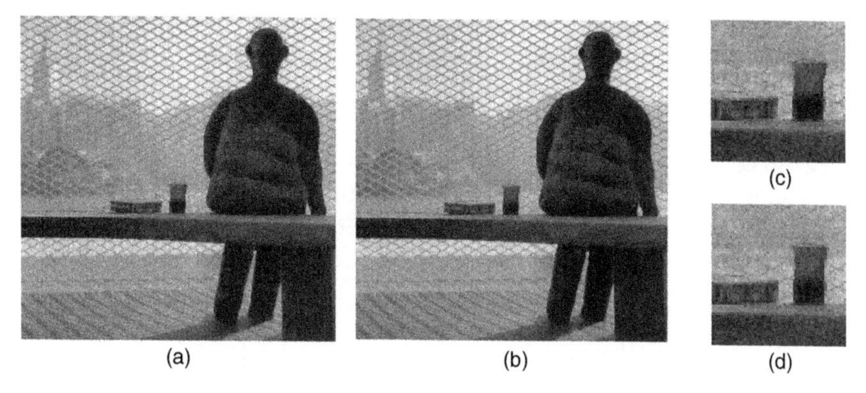

Figure 7.5 Wavelet image fusion denoised image: (a) by fusing average image of Figure 7.3(a) and (b) ($PSNR = 21.0811$ dB), (b) by fusing Figure 7.3(a) and 10 singular value threshold recovered image from Figure 7.3(a) with 50% of the pixel values at random locations being replaced by 0 values, and the random locations of these 10 images has less than 50% correlation ($PSNR = 21.1268$ dB), and (c) and (d) are the zoomed-in portion of the *book and glass* in (a) and (b), respectively.

from the zoomed-in portion of the *glass and book* of Figure 7.5(a), as shown in Figure 7.5(c) that while a lot of noise has been suppressed, the edges of the image objects are well-defined without blurring, the metal meshes are preserved in the denoised image. There are blocking artifacts observed in the denoised image, and the objects with narrow features get enlarged. As an example, while the metal meshes can be preserved in the denoised image, it is vivid that the metal thickness is thicker than that in the noise-free *Sculpture* image.

Let us consider the wavelet fusion of 10 SVT recovered images to determine if the blocking artifacts will be alleviated. All 10 images are stored at `s1` to `s10`. The following MATLAB script will perform the wavelet image fusion for all the 10 recovered images together with the original noisy image f.

```
>> [M,N] = size(s1);
>> ssize = M*N;
>> [w1,j,th] = waveletfus(s1,s2,ssize,0,3);
>> [w2,j,th] = waveletfus(w1,s3,ssize,0,3);
>> [w3,j,th] = waveletfus(w2,s4,ssize,0,3);
>> [w4,j,th] = waveletfus(w3,s5,ssize,0,3);
>> [w5,j,th] = waveletfus(w4,s6,ssize,0,3);
>> [w6,j,th] = waveletfus(w5,s7,ssize,0,3);
>> [w7,j,th] = waveletfus(w6,s8,ssize,0,3);
>> [w8,j,th] = waveletfus(w7,s9,ssize,0,3);
>> [w9,j,th] = waveletfus(w8,s10,ssize,0,3)
>> [g,j,th] = waveletfus(w9,f,ssize,0,3);
```

The image fusion result is shown in Figure 7.5(b), where it is vivid from the zoomed-in portion of the *glass and book* of Figure 7.5(b), as shown in Figure 7.5(d).

It can be observed that the blocking artifacts got alleviated, and the thickening problem of narrow image features is no longer observed in the denoised image. The wavelet fusion effectively removes unwanted high frequency components in the final image thanks to the regularity property of wavelet transform. However, the "ghost edge" problem, which is common in the wavelet-based denoising method, is resulted due to the dislocation problem which degraded the objective performance of the final image where the PSNR is 21.1268 dB, which is lower than 22.4795 dB obtained by the simple averaging of 10 SVT restored images, as shown in Figure 7.4(b). However, the reader should note that its performance is still comparable to those restored from other methods presented in this chapter. The PSNR can be improved by adjusting the number of decomposition levels, the threshold functions, or the threshold values. The "ghost edge" problem hampers the performance on wavelet fusion denoising for AWGN corrupted images, but it might not be the case for mixed noise-corrupted images. Applying wavelet fusion on multiple SVT-restored mixed noise-corrupted images would result in better objective performance when compared to the use of simple averaging of the same set of SVT-restored images because the wavelet-based approach is more effective in removing noise with high sparsity due to the decorrelating property of the wavelet transform.

7.5 Summary

The random sampling method discussed in this chapter sheds a new light on how to introduce randomness to the denoised image set which allows further processing to obtain better denoised image results. The averaging and wavelet fusions are the two processing methods that have discussed in this chapter to process a set of AWGN corrupted image with the same core image to obtain a denoised image. Other methods exist and can be applied, which can provide different image denoising properties to make it practical for your applications. Besides sampling, there are other ways to introduce randomness to assist image denoising. Showing in one of the exercises is the application of random assignment to the NLM filter aiming to achieve better image denoising results.

Exercises

7.1 SVT with different sampling rates. Consider the AWGN corrupted *Sculpture* image with $\sigma_\eta = 10, f$, write the MATLAB programs that achieve the following.

1. Generate another image from f with 10% of the pixels at random locations replaced by 0.
2. Apply SVT to recover the image s from f. Report the PSNR of the recovered image.
3. Apply wavelet fusion with f and s to generate a denoised image. Report the PSNR of the denoised image.

Repeat the above with different percentage of pixels being replaced by 0, and different σ_η of the corrupted *Sculpture* image. Discuss the best percentage of pixels being replaced by 0 that will generate the best-denoised image through fusion with f and if the best percentage depends on the noise power of f. (You may want to plot a 3D surface to sustain your conclusion.)

7.2 Random NLM patch block
Develop a MATLAB function that will process an image noisy image f, similar to Listing 6.1.3 where other inputs to the function include patch size *psize*, the search window size *wsize*, and the smoothing factor h. The MATLAB function will perform the following operations.
For each pixel p_0 in f, find a set of n random pixel locations within the whole image.

1. For each pixel locations, p with the set of n random pixel locations, extract a patch with size $(2 * psize + 1) \times (2 * psize + 1)$.
2. Process each patch window with Equation 6.5 to compute a weight that associates pixel p.
3. After computing all weights for all n pixels, normalize the weights according to Equation 6.2.
4. Compute the weighted average according to Equation 6.1 and use it to replace the intensity of pixel p_0.
5. The denoise image g is obtained when all the pixel intensities in f have been replaced by the associated weighted average.

Try out the developed function with $n = 49$, and $psize = 2$ with AWGN corrupted *Sculpture* image with different σ_η. Report the PSNR of the denoised images. Vary n and try again, and discuss how that affects the performance of the denoised images.

7.3 Create a MATLAB function `svtsamp` to generate Figure 7.4(b).

7.4 Following the same logic as Exercise 4.5, the structural patterned noise in the inpaint images after SVT reconstruction of random sampled noisy images should be able to be alleviated by applying rotation to the image before random sampling and SVT reconstruction, and then rotating the reconstructed image back before averaging. Develop the procedure that will perform this procedure with eight images, where the first image is obtained by rotating the noisy image by 10°, and the second image is obtained by rotating the first image by an additional 10°, and so on. Perform random sampling with the intensity of 50% of the pixel replaced by zero value. Then apply SVT to recover the image. Rotate the image back to the original orientation and then extract the necessary image out from it. Sum up all the eight images into a single image and divide that by 8.

1. Present your MATLAB code, PSNR of the resulting image when the noisy image is constructed with the *Sculpture* image corrupted by AWGN with $\sigma_\eta = 25$, and also plot the figure image.
2. Average the denoised image with the noisy image again. What is the PSNR of the resulting image, and what do you think is the cause of the improvement/decrement in PSNR?

7.5 Compare the performance on denoising the mixed noise-corrupted image presented in Figure 1.11(a) using

1. singular value thresholding inpainting (SVT);
2. averaging of 10 SVT restored images and the original noisy image; and
3. wavelet fusion merging 10 SVT restored images and the original noisy image at level 2 decomposition.

Appendix A

MATLAB Functions List

The following is a list of all the MATLAB sources that are applied to generate the figures and table presenting the simulation results in this book. All MATLAB sources have been tested with MATLAB 2023B. If you are using MATLAB with another version, beware that you may need to modify the sources, as MATLAB do update the language from time to time.

Figure	Listing	Filename	Function
	1.1.1	`readImage`	Digital image details
	1.1.2	`rgb2ycbcr`	RGB to YCbCr conversion
	1.1.3	`ycbcr2rgb`	YCbCr to RGB conversion
	1.1.4	`rgb2gray`	RGB to grayscale image conversion
	1.3.1	`callrgb2cbcr`	MATLAB script calling `rgb2gray`
	1.3.2	`diplayImage`	Display digital image
	1.3.3	`saveImage`	Saving a digital image
	1.3.4	`datacovDouble`	Data type conversion from uint8 to double
	1.3.5	`datacovUint8`	Data type conversion from double to uint8
1.6(a)	1.4.1	`halfPixelExt`	Symmetric extension of image *Sculpture*
	1.4.2	`wholePixelExt`	Whole pixel symmetric extension by MATLAB built-in function `wextend`
	1.5.1	`awgnNoise`	MATLAB script generating AWGN corrupted image
	1.5.2	`db2linear`	MATLAB script converting dB scale to linear scale

Digital Image Denoising in MATLAB, First Edition. Chi-Wah Kok and Wing-Shan Tam.

Companion website: www.wiley.com/go/kokDeNoise

Figure	Listing	Filename	Function
	1.5.3	`awgnPredfinedSNR`	MATLAB script generating AWGN corrupted image with predetermined SNR
	1.5.4	`imgtrim`	Pixel values truncation to the range of `uint8`, while maintaining the data type
1.8(a)		*`sculpture`*`AWGN10`	AWGN corrupted *Sculpture* image with zero mean and $\sigma_\eta = 10$
1.8(b)		*`sculpture`*`AWGN50`	AWGN corrupted *Sculpture* image with zero mean and $\sigma_\eta = 50$
1.8(c)		`uniformAWGN10`	AWGN corrupted uniform tone image with intensity level at 128 and $\sigma_\eta = 10$
1.9(a)	1.5.5	`nestls`	Noise estimation by local statistic on the uniform tone image corrupted by AWGN with $\sigma_\eta = 50$
1.9(a)	1.5.6	`nhist`	Histogram of the noise variance computed from 7 × 7 mask of uniform tone image corrupted by AWGN with $\sigma_\eta = 50$
1.9(b)	1.5.5	`nestls`	Noise estimation by local statistic on the *Sculpture* image corrupted by AWGN with $\sigma_\eta = 50$
1.9(b)	1.5.6	`nhist`	Histogram of the noise variance computed from 7 × 7 mask of *Sculpture* image corrupted by AWGN with $\sigma_\eta = 50$
	1.5.7	`deriv`	2D derivative
	1.5.8	`derisigmaest`	Noise estimate via derivative
	1.5.9	`sapnoise`	Salt and pepper noise (SAP)
1.10(a)		`uniformSAP`	SAP corrupted uniform tone image with intensity level at 128 and total noise density at 0.05
1.10(b)		*`sculpture`*`SAP`	SAP corrupted *Sculpture* image with total noise density at 0.05
1.11(a)	1.6.1	`mixNoiseAWGNthenSAP`	*Sculpture* image corrupted with AWGN with $\sigma_\eta = 50$, then SAP with density 0.05

Figure	Listing	Filename	Function
1.11(b)	1.6.2	`mixNoiseSAPthenAWGN`	*Sculpture* image corrupted with SAP with density 0.05, then AWGN with $\sigma_\eta = 50$
	1.8.1	`imageerr`	Error image
	1.8.2	`mae`	Mean absolute error
	1.8.3	`mse`	Mean squares error
	1.8.4	`rmse`	Root mean squares error
	1.8.5	`psnr`	Peak signal-to-noise ratio (PSNR)
1.14(a)		`sculptureAWGN50`	*Sculpture* image corrupted by AWGN uniformly across the whole image with $\sigma_\eta = 50$
1.14(b)	1.8.6	`imageSpatialNoise`	*Sculpture* image corrupted by AWGN in texture area alone with $\sigma_\eta = 2.59$
1.14(c)	1.8.6	`imageSpatialNoise`	*Sculpture* image corrupted by AWGN in flat area alone with $\sigma_\eta = 1.47$
	1.8.7	`fpsnr`	Peak signal-to-noise ratio on flat region
	1.8.8	`tpsnr`	Peak signal-to-noise ratio on texture region
	1.8.9	`sobel`	Sobel edge image extraction
	1.8.10	`tmap`	Texture map (tmap)
	1.8.11	`epsnr`	Edge peak signal-to-noise ratio (EPSNR)
	1.9.1	`mssim`	Mean structural similarity index
	1.10.1	`brightnorm`	Brightness normalization
2.3(a)	2.1.1	`idealFilter`	512×512 ideal filter in 2D view in frequency domain with $r_0 = 50$
2.3(b)	2.1.2	`idealFilterDenoise`	The denoised *Sculpture* image using ideal lowpass filter with image corrupted with AWGN having $\sigma_\eta = 50$
	2.1.3	`meanFilterBuiltin`	MATLAB built-in function for mean filtering
2.4(a)		`meanFilterFreq`	Mean filter in frequency domain
2.4(b)	2.1.3	`meanFilterBuiltin`	The denoised *Sculpture* image using a 3×3 mean filter kernal applied to AWGN corrupted image having $\sigma_\eta = 50$ using MATLAB built-in function for mean filtering

Figure	Listing	Filename	Function
	2.1.4	`meanFilter3`	The denoised *Sculpture* image using a 3×3 mean filter kernal applied to AWGN corrupted image having $\sigma_\eta = 50$
2.5(a)		`meanFilterDenoise`	Effect of mean filter kernel size versus different sources of noise – 5×5 mean filter on AWGN corrupted *Sculpture* image with $\sigma_\eta = 50$
2.5(b)		`meanFilterDenoise`	Effect of mean filter kernel size versus different sources of noise – 5×5 mean filter on AWGN and SAP mixed noise-corrupted *Sculpture* image
2.5(c)		`meanFilterDenoise`	Effect of mean filter kernel size versus different sources of noise – 3×3 mean filter on AWGN and SAP mixed noise-corrupted *Sculpture* image
2.6(a)	2.1.5	`gaussianFilter`	Spatial frequency response of Gaussian filter along m-axis with kernel size 5×5
2.6(b)	2.1.5	`gaussianFilter`	3D view of the frequency response of the Gaussian filter in Figure 2.6(a)
2.6(c)	2.1.5	`gaussianFilter`	Aerial view of the frequency response of the Gaussian filter in Figure 2.6(b)
	2.1.6	`gaussiankernel`	Gaussian filter kernel generation
2.7(a)		`gaussianFilterDenoise`	Effect of the size of a Gaussian filter with identical passband roll off factor $\sigma = 5$ versus different sources of noise – 5×5 Gaussian filter applied to AWGN corrupted image
2.7(b)		`gaussianFilterDenoise`	Effect of the size of a Gaussian filter with identical passband roll off factor $\sigma = 5$ versus different sources of noise – 5×5 Gaussian filter applied to AWGN and SAP mixed noise-corrupted image
2.7(c)		`gaussianFilterDenoise`	Effect of the size of a Gaussian filter with identical passband roll off factor $\sigma = 5$ versus different sources of noise – 3×3 Gaussian filter applied to AWGN corrupted image

Figure	Listing	Filename	Function
2.7(d)		gaussianFilterDenoise	Effect of the size of a Gaussian filter with identical passband roll off factor $\sigma = 5$ versus different sources of noise – 3×3 Gaussian filter applied to AWGN and SAP mixed noise-corrupted image
2.8(a)		wienerFilterDenoise	Wiener filtering denoised *Sculpture* image corrupted with AWGN having $\sigma_\eta = 10$
2.8(b)		wienerFilterDenoise	Wiener filtering denoised *Sculpture* image corrupted with AWGN having $\sigma_\eta = 50$
2.8(c)		wienerFilterDenoise	Wiener filtering denoised *Sculpture* image corrupted with AWGN and SAP mixed noise
	2.3.1	blockdct	Forward $L \times L$ DCT block processing of image array f
	2.3.2	dctBlock8	Forward 8×8 DCT block processing of image array f
	2.3.3	blockidct	Inverse $L \times L$ DCT block process of image array f
	2.3.4	hthfun	Hard thresholding
2.9(a)	2.3.5	htdct8	AWGN corrupted *Sculpture* image denoised by 8×8 non-overlap blocked DCT hard threshold denoising
2.9(b)	2.3.6	htdctHalf	AWGN corrupted *Sculpture* image denoised by 8×8 half-shifted blocked DCT hard threshold denoising
2.9(b)	2.3.7	htdctHalfOverlap	AWGN corrupted *Sculpture* image denoised by averaging the result images of 8×8 non-overlap blocked DCT hard threshold denoising and 8×8 half-shifted block DCT hard threshold denoising
	2.4.1	medianfilter	Median filtering
2.11(a)	2.4.2	medianfilterDenoise	Image denoising by median filtering on SAP corrupted *Sculpture* image with a density of 0.05

Figure	Listing	Filename	Function
2.11(b)	2.4.2	medianfilterDenoise	Image denoising by median filtering on AWGN and SAP mixed noise-corrupted *Sculpture* image
	2.4.3	medianfilteradp	Median filtering with adaptive window
2.12(a)	2.4.4	medianfilteradpDenoise	Image denoising by median filtering with adaptive window size ranging from 3 × 3 to 5 × 5 on SAP corrupted *Sculpture* image
2.12(b)	2.4.4	medianfilteradpDenoise	Image denoising by median filtering with adaptive window size ranging from 3 × 3 to 5 × 5 on AWGN and SAP mixed-noise corrupted *Sculpture* image
	2.4.5	medianfiltermask	Median filtering with predefined mask
2.13(a)		medianfiltermaskDenoise	Image denoising by median filtering with mask medneigh on SAP noise corrupted *Sculpture* image
2.13(b)		medianfiltermaskDenoise	Image denoising by median filtering with mask medneigh on AWGN and SAP mixed noise-corrupted *Sculpture* image
	2.4.6	medianmedian	Median of median filtering
2.14(a)	2.4.6	medianmedianDenoise	Image denoising by median of median filtering on SAP corrupted *Sculpture* image
2.14(b)	2.4.6	medianmedianDenoise	Image denoising by median of median filtering on AWGN and SAP mixed noise-corrupted *Sculpture* image
	3.1.1	singleLevelDWT	Single-level discrete Haar wavelet transform
	3.1.2	twoLevelDWT	Two-level discrete Haar wavelet transform
3.1(c)	3.1.3	displayDWTD	Display of single-level discrete Haar wavelet transform image
3.1(e)	3.1.3	displayDWTD	Display of two-level discrete Haar wavelet transform image
	3.1.4	inverseDWTD	Inverse discrete wavelet transform
	3.2.1	waveletnoiseest	Wavelet domain image noise estimation

Figure	Listing	Filename	Function
3.2	3.2.2	histogramWaveletnoise	Plot of the histogram of fhh3 wavelet coefficients
3.3(a)–(c)	3.3.1	sculptureColDWT	1D discrete wavelet transform of selected column in *Sculpture* image
3.3(d)–(f)		sculptureAWGNColDWT	1D discrete wavelet transform of selected column in the AWGN corrupted *Sculpture* image
	3.4.1	waveletth	Wavelet threshold denoising
	3.4.2	hthfun	Hard thresholding
	3.4.3	sthfun	Soft thresholding
	3.4.4	gthfun	Garrote thresholding
3.6	3.5.1	msevsth	MSE performance of different wavelet thresholding functions with varying threshold values
3.6		wavelethvar	Wavelet thresholding function with varying hard threshold values
3.6		waveletsthvar	Wavelet thresholding function with varying soft threshold values
3.6		waveletgthvar	Wavelet thresholding function with varying Garrote threshold values
	3.5.2	thest	Threshold value estimation function for wavelet thresholding
3.7(a)	3.5.3	whichfun	Wavelet threshold image denoising with hard threshold on *Sculpture* image corrupted with AWGN having $\sigma_\eta = 25$
3.7(a)	3.5.4	thfun	Wavelet threshold function selector
3.7(b)	3.5.3	whichfun	Wavelet threshold image denoising with soft threshold on *Sculpture* image corrupted with AWGN having $\sigma_\eta = 25$
3.7(b)	3.5.4	thfun	Wavelet threshold function selector
3.7(c)	3.5.3	whichfun	Wavelet threshold image denoising with Garrote threshold on *Sculpture* image corrupted with AWGN having $\sigma_\eta = 25$
3.7(c)	3.5.4	thfun	Wavelet threshold function selector

Figure	Listing	Filename	Function
3.8(a)		waveletDenoise	MATLAB script calling waveletth
3.8(a)	3.4.1	waveletth	Wavelet hard threshold denoising without adaptive threshold on *Sculpture* image corrupted with AWGN having $\sigma_\eta = 25$
3.8(b)		waveletDenoise	MATLAB script calling waveletath
3.8(b)	3.5.5	waveletath	Wavelet hard threshold denoising with adaptive threshold on *Sculpture* image corrupted with AWGN having $\sigma_\eta = 25$
3.8(c)		waveletDenoise	MATLAB script calling waveletsth
3.8(c)	3.5.6	waveletsth	Wavelet hard threshold denoising with scale shrink threshold on *Sculpture* image corrupted with AWGN having $\sigma_\eta = 25$
3.9(a)		waveletwienerDenoise	MATLAB script calling waveletwiener
3.9(a)	3.6.1	waveletwiener	Image denoising combining wavelet hard threshold denoising with scale shrink threshold and Wiener filtering on *Sculpture* image corrupted with AWGN having $\sigma_\eta = 25$
3.9(b)		wienerFilterDenoise	Image denoising by Wiener filtering with filter size of 3×3 on *Sculpture* image corrupted with AWGN having $\sigma_\eta = 25$
3.11	3.7.2	cyclespinDenoise	The image quality against the number of translation (spinsize) on the wavelet hard threshold denoising with scale shrink threshold
3.11	3.7.1	cyclespin	Wavelet hard threshold denoising with scale shrink threshold and cycle spinning
3.12	3.7.3	cyclespinExample	Wavelet hard threshold denoising with scale shrink threshold and seven cycle spinnings
3.12	3.7.1	cyclespin	Wavelet hard threshold denoising with scale shrink threshold and cycle spinning
	3.8.1	waveletfusDenoise	MATLAB script calling waveletfus

Figure	Listing	Filename	Function
3.14(a)		waveletfus2imgDenoise	Input image applied in wavelet fusion in denoising *Sculpture* image corrupted with AWGN having $\sigma_\eta = 25$
3.14(b)	3.8.2	waveletfus	Main function of Wavelet based image fusion denoising
3.14(b)	3.8.3	maxmag	Function returning a matrix containing the entry from the same location of two matrices having the greater magnitude
3.14(b)		waveletfus2imgDenoise	Image denoising by wavelet fusion on *Sculpture* image corrupted with AWGN having $\sigma_\eta = 25$
	4.1.1	svdImage	SVD an image
4.1(a)–(c)	4.1.2	plotsvdimage	Plot of the squares of singular values of an image
	4.2.1	svdsigmaest	Estimating σ_η by Equation 4.15 via SVD
	4.2.2	svdsigmamad	Estimating σ_η by Equation 1.14 via SVD
	4.2.3	svdEstimateNoise	MATLAB script calling svdsigmaest and svdsigmamad
4.2(a)	4.2.4	svdHardThreshold	Image denoising by SVD hard thresholding with shrinker threshold
4.2(b)	4.2.4	svdHardThreshold	Image denoising by SVD hard thresholding with optimal threshold
4.2(c)	4.2.5	svdHardThresholdsigma	Image denoising by SVD hard thresholding using svdsigmamad to estimate σ_η
	4.3.1	blocksvd	Block SVD image denoised by hard threshold and estimated σ_η
	4.3.2	svdhardth	SVD hard threshold with estimated σ_η
4.3(a)	4.3.3	blocksvdDenoise	32×32 non-overlap block SVD image denoising by hard threshold and estimated σ_η on AWGN corrupted *Sculpture* image having $\sigma_\eta = 50$

Figure	Listing	Filename	Function
4.3(b)	4.3.4	averageBlocksvdDenoise	Average of two non-overlap block SVD image denoising obtained by hard threshold and estimated σ_η, where the block of one image is half-shifted when compared to the other on AWGN corrupted *Sculpture* image having $\sigma_\eta = 50$
4.4(a)		svdhardth	Non-blockwise SVD image denoising by hard thresholding using svdsigmamad to estimate σ_η on AWGN and SAP mixed noise-corrupted *Sculpture* image
4.4(b)		blocksvd32nonoveralp	32 × 32 non-overlap block SVD image denoising by hard threshold and estimated σ_η on AWGN and SAP mixed noise-corrupted *Sculpture* image
4.4(c)		blocksvd16nonoverlap	16 × 16 non-overlap block SVD image denoising by hard threshold and estimated σ_η on AWGN and SAP mixed noise-corrupted *Sculpture* image
	4.4.1	rsvd	Randomized SVD (RSVD)
4.5(a)	4.4.2	rsvdDenoise	Image denoising by RSVD on AWGN corrupted *Sculpture* image with $\sigma_\eta = 50$
4.5(b)	4.4.2	rsvdDenoise	Image denoising by RSVD on AWGN and SAP mixed noise-corrupted *Sculpture* image
4.6(a)	4.4.3	irsvd	Image denoising by iterative RSVD on AWGN corrupted *Sculpture* image with $\sigma_\eta = 50$
4.6(b)	4.4.3	irsvd	Image denoising by iterative RSVD on AWGN and SAP mixed noise-corrupted *Sculpture* image
	5.2.1	roftv	ROF total variation image denoising
5.1(a)	5.2.2	roftvDenoise	ROF total variation image denoising on AWGN corrupted *Sculpture* image with $\sigma_\eta = 50$ using $\lambda = 0.01$

Figure	Listing	Filename	Function
5.1(b)	5.2.2	`roftvDenoise`	ROF total variation image denoising on AWGN corrupted *Sculpture* image with $\sigma_\eta = 50$ using $\lambda = 0.05$
5.1(c)	5.2.2	`roftvDenoise`	ROF total variation image denoising on AWGN corrupted *Sculpture* image with $\sigma_\eta = 50$ using $\lambda = 0.1$
	6.1.1	`patchDiff`	MATLAB script for computing the weightings between the local patch and the neighboring patch
	6.1.2	`nlmFramework`	NonLocal means filtering framework
	6.1.3	`nlm`	NonLocal means filtering
6.3(a)		`nlmDenoise`	Image denoising by NonLocal means filtering on AWGN corrupted *Sculpture* image with $\sigma_\eta = 25$
6.3(b)		`nlmDenoise`	Image denoising by NonLocal means filtering on AWGN and SAP mixed noise-corrupted *Sculpture* image
	6.1.4	`nlmhth`	NonLocal means filtering with hard threshold
6.5(a)		`nlmthDenoise`	Image denoising by NonLocal means filtering with hard threshold on AWGN corrupted *Sculpture* image with $\sigma_\eta = 25$
6.5(b)		`nlmthDenoise`	Image denoising by NonLocal means filtering with hard threshold on AWGN and SAP mixed noise-corrupted *Sculpture* image
6.6		`nlmPSNR`	PSNR performance of NonLocal means filtering on AWGN corrupted *Sculpture* image with $\sigma_\eta = 25$ with respect to different patch sizes and search window sizes
	6.2.1	`nlmadp`	NonLocal means filtering with adaptive patch window size
6.7(a)		`nlmadpDenoise`	Image denoising by NonLocal means filtering with adaptive patch size on AWGN corrupted *Sculpture* image with $\sigma_\eta = 25$
6.7(b)		`nlmadpDenoise`	Image denoising by NonLocal means filtering with adaptive patch size on AWGN and SAP mixed noise-corrupted *Sculpture* image
	6.2.2	`nlmadpw`	NonLocal means filtering with adaptive search window size

Figure	Listing	Filename	Function
6.8(a)		`nlmadpwDenoise`	Image denoising by NonLocal means filtering with adaptive search window size on AWGN corrupted *Sculpture* image with $\sigma_\eta = 25$
6.8(b)		`nlmadpwDenoise`	Image denoising by NonLocal means filtering with adaptive search window size on AWGN and SAP mixed noise-corrupted *Sculpture* image
7.2(a)	7.1.1	`averageMultipleImage`	One of the ten AWGN corrupted *Sculpture* images with $\sigma_\eta = 50$
7.2(b)	7.1.1	`averageMultipleImage`	The average of all ten AWGN corrupted *Sculpture* images with $\sigma_\eta = 50$
	7.3.1	`svt`	Singular value thresholding inpainting (SVT)
7.3(a)		*`sculpture`*`AWGN25`	AWGN corrupted *Sculpture* image with $\sigma_\eta = 25$
7.3(b)		`randsamp`	Intermediate image generated from AWGN corrupted *Sculpture* image with $\sigma_\eta = 25$ having 50% of its pixels assigned to zero at random locations
7.3(c)		`svtfill`	SVT recovered image from the intermediate image
7.5(a)		`randsampwfuse`	Wavelet based image fusion denoising by fusing the average image of the original noisy image and the intermediate image generated from AWGN corrupted *Sculpture* image with $\sigma_\eta = 25$ having 50% of its pixels assigned to zero at random locations
7.5(b)		`randsampwfuse10`	Wavelet based image fusion denoising by fusing the original noisy image and 10 SVT recovered images

Table	Listing	Filename	Function
3.2		`compuWaveletCoeff`	Computing of the properties of `fhh3` wavelet coefficient distribution

References

1 K.J. Arrow, L. Hurwicz, and H. Uzawa. Iterative methods for concave programming. *Studies in linear and nonlinear programming*, 1958.

2 G. Aubert and J. Aujol. A variational approach to remove multiplicative noise. *SIAM Journal on Applied Mathematics*, 68(4):925–946, 2008.

3 B.E. Bayer. Color imaging array. *US Patent No. 3971065*, 1976.

4 A. Beck and M. Teboulle. A fast iterative shrinkage-thresholding algorithm for linear inverse problems. *SIAM Journal on Imaging Science*, 2:183–202, 2009.

5 A. Buades, B. Coll, and J.-M. Morel. A non-local algorithm for image denoising. *Proceedings of the 2005 IEEE Computer Society Conference on Computer Vision and Pattern Recognition (CVPR'05)*, Volume 2, pages 60–65, 2005.

6 P. Burt and R. Kolczynski. Enhanced image capture through fusion. *Proceedings of the 4th International Conference on Computer Vision*, pages 173–182, 1993.

7 J.F. Cai, E.J. Candes, and Z. Shen. A singular value thresholding algorithm for matrix completion. *SIAM Journal on Optimization*, 20(4):1956–1982, 2010.

8 V. Caselles, G. Sapiro, and D.H. Chung. Vector median filters, inf-sup operations, and coupled PDE's: theoretical connections. *Journal of Mathematical Imaging and Vision*, 12(3):109–119, 2000.

9 A. Chambolle. An algorithm for total variation minimization and applications. *Journal of Mathematical Imaging and Vision*, 20(1):89–97, 2004.

10 A. Chambolle and J. Darbon. On total variation minimization and surface evolution using parametric maximum flows. *International Journal of Computer Vision*, 84(3):288–307, 2009.

11 S.G. Chang, B. Yu, and M. Vetterli. Adaptive wavelet thresholding for image denoising and compression. *IEEE Transactions on Image Processing*, 9:1532–1546, 2000.

12 P. Chatterjee and P. Milanfar. Is denoising dead? *IEEE Transactions on Image Processing*, 19(4):895–911, 2010.

Digital Image Denoising in MATLAB, First Edition. Chi-Wah Kok and Wing-Shan Tam.

Companion website: www.wiley.com/go/kokDeNoise

13 R.R. Coifman and D.L. Donoho. Translation invariant de-noising. In *Lecture Notes in Statistics: Wavelets and Statistics*, volume 103, pages 125–150. Springer-Verlag, 1995.

14 M.A. Davenport and J. Romberg. An overview of low-rank matrix recovery from incomplete observations. *IEEE Journal of Selected Topics in Signal Processing*, 10(4):608–622, 2016.

15 C.A. Deledalle, V. Duval, and J. Salmon. Non-local methods with shape-adaptive patches (NLM-SAP). *Journal of Mathematical Imaging and Vision*, 43(2):103–120, 2012.

16 D.L. Donoho. Wavelet thresholding and W.V.D.: a 10-minute tour. *Proceeding of the International Conference on Wavelets and Applications*, 1992.

17 D.L. Donoho. De-noising by soft-thresholding. *IEEE Transactions on Information Theory*, 41(3):613–627, 1995.

18 D.L. Donoho and J.M. Johnstone. Ideal spatial adaptation by wavele shrinkage. *Biometrika*, 81(3):425–455, 1994.

19 C. Eckart and G. Young. The approximation of one matrix by another of lower rank. *Psychometrika*, 1(3):211–218, 1936.

20 M.P. Eckert and A.P. Bradley. Perceptual quality metrics applied to still image compression. *Signal Processing*, 70(11):177–200, 1998.

21 M. Fazel, H. Hindi, and S.P. Boyd. A rank minimization heuristic with application to minimum order system approximation. *Proceedings of the American Control Conference*, pages 4734–4739, 2001.

22 J.A. Ferwerda, S.N. Pattanaik, P. Shirley, and D.P. Greenberg. A model of visual masking for computer graphics. *ACM SIGGRAPH '97 Conference Proceedings of Computer Graphics*, pages 143–152, 1997.

23 M.A.T. Figueiredo and R.D. Nowak. Wavelet-based image estimation: an empirical Bayes approach using Jeffrey's noninformative prior. *IEEE Transaction on Image Processing*, 10(9):1322–1331, 2001.

24 H.Y. Gao. Wavelet shrinkage denoising using the non-negative garrote. *Journal of Computer Graphics and Statistics*, 7:469–488, 1998.

25 M. Gavish and D.L. Donoho. The optimal hard threshold for singular values is 4/sqrt(3). *IEEE Transactions on Information Theory*, 60(8):6040–5053, 2014.

26 R.C. Gonzalez and R.W. Woods. *Digital Image Processing*. Prentice-Hall, 2002.

27 N. Halko, P.G. Martinsson, and J.A. Tropp. Finding structure with randomness: probabilistic algorithms for constructing approximate matrix decompositions. *SIAM Review*, 53:217–288, 2011.

28 D.M. Hawkins, L. Liu, and S.S. Young. Robust singular value decomposition. *NISTR*, 122, 2001.

29 J. Hu and Y.P. Luo. Non-local means algorithm with adaptive patch size and bandwidth. *Optik-International Journal for Light and Electron Optics*, 124(22):5639–5645, 2013.

30 M. Jansen. *Noise Reduction by Wavelet Thresholding*, volume 161. Springer, 2001.

31 W.B. Johnson and J. Lindenstrauss. Extensions of Lipschitz mappings into a Hilbert space. *Contemporary Mathematics*, 26(1):189–206, 1984.

32 C.-W. Kok and W.-S. Tam. *Digital Image Interpolation in MATLAB*. Wiley-IEEE, 2019.

33 T. Le, R. Chartrang, and T.J. Asaki. A variational approach to reconstructing images corrupted by Poisson noise. *Journal of Mathematical Imaging and Vision*, 27(3):257–263, 2007.

34 X. Li and M.T. Orchard. New edge-directed interpolation. *IEEE Transactions on Image Processing*, 10(10):1521–1527, 2001.

35 H. Li, B. Manjunath, and S. Mitra. Multisensor image fusion using the wavelet transform. *Graphical, Models and Image Processing*, 57(3):235–245, 1995.

36 V.V. Lukin, R. Oktem, N. Ponomarenko, and K. Egiazarian. Image filtering based on discrete cosine transform. *Journal of Telecommunication Radio Engineering*, 66(18):1685–1701, 2007.

37 V. Lukin, N. Ponomarenko, and K. Egiazarian. HVS-metric-based performance analysis of image denoising algorithms. *Proceedings of the EUVIP*, pages 156–161, 2011.

38 S.G. Mallat. *A Wavelet Tour of Signal Processing*. Academic Press, 1998.

39 M.K. Ng, L. Qi, Y.F. Yang, and Y.M. Huang. On semismooth Newton's methods for total variation minimization. *Journal of Mathematical Imaging and Vision*, 27(3):265–276, 2007.

40 W.B. Pennebaker and J.L. Mitchell. *JPEG Still Image Data Compression Standard*. Springer, 3rd edition, 1993.

41 V. Petrovic and C. Xydeas. Graident based multiresolution image fusion. *IEEE Transactions on Image Processing*, 13(2):228–237, 2004.

42 O. Pogrebnyak and V. Lukin. Wiener discrete cosine transform-based image filtering. *Journal of Electronic Imaging*, 21(4):043020 2012.

43 J.G. Proakis and M. Salehi. *Digital Communications*. McGraw-Hill Education, 5th edition, 2007.

44 Recommendation ITU-R BT.601-5. Studio encoding parameters of digital television for standard 4:3 and wide-screen 16:9 aspect ratios. *ITU-T*, 1995.

45 Recommendation J.144 (03/04). Objective perceptual video quality measurement techniques for digital cable television in the presence of a full reference. *International Telecommunication Union, Telecommunication Standardization Sector*, March 2001.

46 L.I. Rudin, S. Osher, and E. Fatemi. Nonlinear total variation based noise removal algorithms. *Physica D*, 60:259–268, 1992.

47 S. Sreejith and J. Nayak. Study of hybrid median filter for the removal of various noises in digital image. *Journal of Physics: Conference Series*, 1706:012079 2020.

48 G. Strang and T.Q. Nguyen. *Wavelets and Filter Banks*. Wellesley-Cambridge Press, 1998.

49 W. Sun and M. Han. Adaptive search based non-local means image de-noising. *2nd International Congress on Image and Signal Processing, CISP'09*, 2009.

50 Z. Wang, A.C. Bovik, H.R. Sheikh, and E.P. Simonelli. Image quality assessment: from error visibility to structural similarity. *IEEE Transactions on Image Processing*, 13(4):600–612, 2004.

51 S. Winkler. Issues in vision modeling for perceptual video quality assessment. *Signal Processing*, 78:231–252, 1999.

52 G. Wyszecki and W. S. Stiles. *Color Science*. John Wiley & Sons, 1982.

53 L. Yaroslavsky and E. Eden. *Fundamentals of Digital Optics: Digital Signal Processing in Optics and Holography*. Birkhauser, 1995.

54 W.L. Zeng and X.B. Lu. Region-based non-local means algorithm for noise removal. *Electronics Letters*, 47(20):1125–1127, 2011.

55 Y. Zhu and C. Huang. An improved median filtering algorithm for image noise reduction. *Physics Procedia*, 25:609–616, 2012.

Index

Digital Image Denoising in MATLAB, First Edition. Chi-Wah Kok and Wing-Shan Tam.

Companion website: www.wiley.com/go/kokDeNoise